AF294360

MODELING OF INDUCTION MOTORS WITH ONE AND TWO DEGREES OF MECHANICAL FREEDOM

MODELING OF INDUCTION MOTORS WITH ONE AND TWO DEGREES OF MECHANICAL FREEDOM

Edited by

ERNEST MENDRELA
Louisiana State University

JANINA FLESZAR and **EWA GIERCZAK**
Kielce University of Technology

Springer Science+Business Media, LLC

Library of Congress Cataloging-in-Publication Data

A C.I.P. Catalogue record for this book is available
from the Library of Congress.

Modeling of induction motors with one and two degrees of mechanical freedom /
  edited by Ernest Mendrela, Janina Fleszar, Ewa Gierczak.
    p. cm
    ISBN 978-1-4613-5055-2   ISBN 978-1-4615-0395-8 (eBook)
    DOI 10.1007/978-1-4615-0395-8
    1. Electric motors, Induction--Mathematical models. I. Mendrela, Ernest. II. Fleszar,
Janina. III. Gierczak, Ewa.

TK2785.M58 2003
621.46--dc21

                                                                2003056510

Copyright © 2003 by Springer Science+Business Media New York
          Originally published by Kluwer Academic Publishers in 2003
          Softcover reprint of the hardcover 1st edition 2003

All rights reserved. No part of this work may be reproduced, stored in a retrieval
system, or transmitted in any form or by any means, electronic, mechanical,
photocopying, microfilming, recording, or otherwise, without the written permission
from the Publisher, with the exception of any material supplied specifically for the
purpose of being entered and executed on a computer system, for exclusive use by
the purchaser of the work.

Printed on acid-free paper.

Contents

Preface

To design electric machines or to study their performance, different types of mathematical models are applied. Some of them are formulated in the circuit theory, others are described on the basis of the field theory. The models, which are used in preliminary design calculations, are based on circuit theory. To optimize the constructional parameters field models are recommended. In the analysis of machine performance circuit models are most frequently used. The parameters of the equivalent circuit model are determined then on the basis of the field model.

The circuit models are most often used in the analysis of induction machines performance when they are supplied from steady voltage source. Applying Fourier series technique to solve differential equations both time and space harmonics can then be considered. This type of approach is used in this book to study the performance of induction motors with one and two degrees of mechanical freedom. First the motors are described in field theory, then to study their behavior with a steady voltage supply, the circuit model is applied, whose parameters are determined from electromagnetic field equations. The advantage of using the Fourier series technique in a description of the field model is, that it allows to pass directly to the multi-harmonic equivalent circuit, and to analyze the motor behavior in steady or dynamic conditions with the inclusion of all the effects taking place at the motor edges. This direct transition from the field model to the circuit model cannot be done when numerical methods are applied.

The authors formulate the model for the motor with two degrees of mechanical freedom. Such a model is more general than the one for conventional rotary motors. It allows the description of sometimes very complex phenomena that take place not only in motors with two degrees of mechanical freedom but also those, which are linked with rotary and linear machines e.g. disc motors, linear motors, liquid metal pumps.

The authors present exclusively their own material developed for more than twenty years and published in a number of journals and conference proceedings. The first simplified model has been formulated for the rotary-linear motor by the author in the middle of 1970's. The theory, based on the magnetic field wave moving helically was developed and extended later to the analysis of X-Y flat linear motor and motors with a spherical rotor. The model formulated in 1980's allowed the study of all phenomena linked with finite length and width of linear and rotary-linear motors. In 1990's the model was applied to analyze the performance of liquid metal pumps and disc induction motors. The results obtained from simulations are backed by the measurements carried out on the motor prototypes of rotary and linear motors as well as on rotary-linear ones. This allowed verifying the mathematical models used for motor analysis.

The book may be found useful for researchers and electrical engineering students, in particular for those who design and analyze induction machines for special purposes.

Acknowledgements

The authors are particular grateful to Mr. Wiesław Muszczak who manufactured the unique measurement stand and all the rotary-linear machines.

Chapter 1

INTRODUCTION

Induction motors are the most commonly applied machines in electric drives. Most of them are squirrel cage motors, mainly due to the easy and low cost technology. Though, there is a large group of linear motors, rotating motors are predominant among induction machines.

Both, rotating and linear motors are classified as machines with one degree of mechanical freedom, since their moving part only runs in one direction. In general, machines with one degree of mechanical freedom have two types of geometry: cylindrical and flat. Most of the rotating machines have the cylindrical shape. The disc motors are rotating machines with a flat structure. Among linear motors both flat and cylindrical geometries are met. These latter are called tubular motors.

There is a group of motors whose moving part is running in the direction that varies during operation e.g. X-Y motor. The motion of these motors must be described in a system with two space co-ordinates. These machines form a group of motors with two degrees of mechanical freedom. Considering the geometry, three types of motors are distinguished:
– X-Y motors – flat structure,
– Rotary-linear motors – cylindrical geometry, and
– Motors with spherical rotors – spherical geometry.

The representative of X-Y motor is shown schematically in Fig. 1-1. The armature (primary part) contains two 3-phase windings placed perpendicularly to one another. If supplied from 3-phase source they produce magnetic traveling fields, which are moving in two perpendicular directions. A secondary part consists of aluminum or copper sheet backed by an iron plate. The forces produced by two traveling fields act in the directions of the field motion. Their values can be changed independently, what contributes to the change of the magnitude and the direction of the resultant force. This in turn influences the motion direction of the X-Y motor.

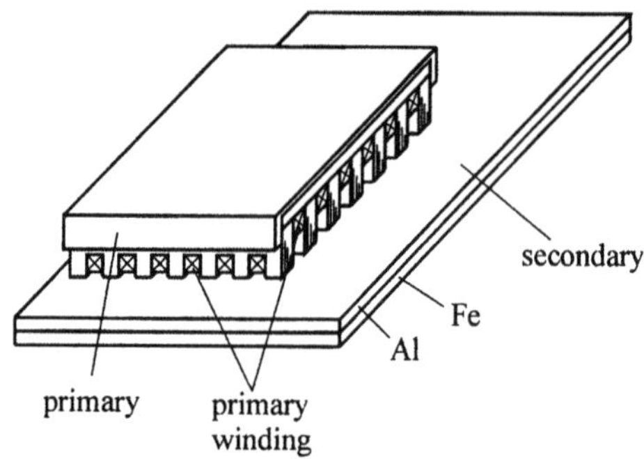

Figure 1-1. Construction scheme of X-Y induction motor

One of a few rotary-linear induction motors is shown in Fig. 1-2. It is a twin-armature rotary-linear motor. Its stator consists of two armatures. One of them, similar to a rotary induction motor stator, produces a rotating field, a second one, similar to a tubular linear induction motor armature generates a field traveling axially. The rotor, common to these two armatures, consists of a steel cylinder covered by a conducting sheet. The direction of the rotor motion depends on two forces: linear and rotary (tangential), which are the products of two magnetic fields and currents induced in. The operation of the motor can be considered as the work of two independent motors: linear and rotary whose rotors are coupled firmly by a mechanical clutch.

The representation of the induction motor with spherical rotor is shown in Fig. 1-3. It is the twin-armature motor whose spherical rotor is driven by

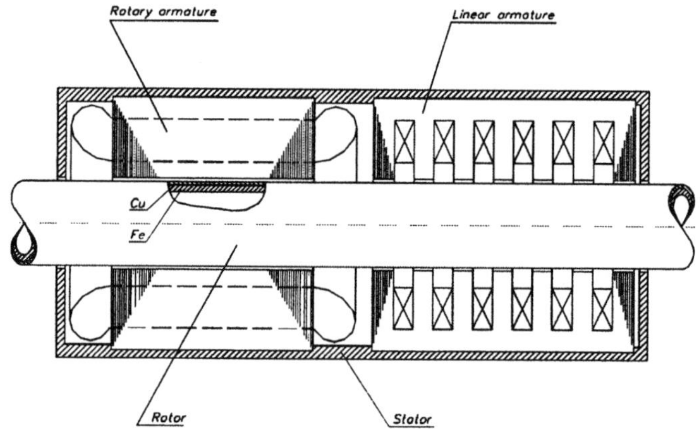

Figure 1-2. Scheme of twin-armature rotary-linear induction motor

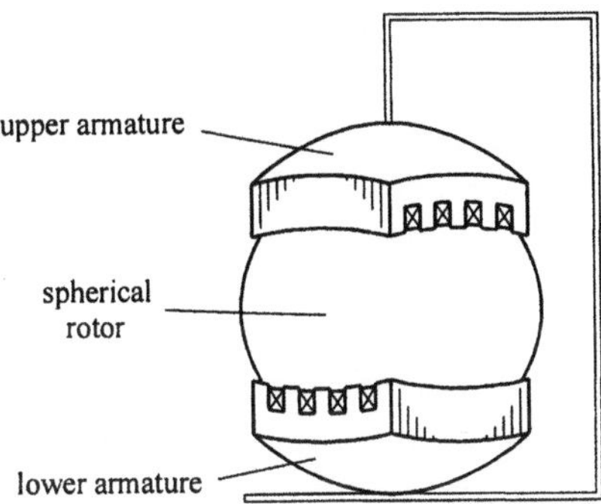

Figure 1-3. Construction scheme of twin-armature induction motor with spherical rotor

two magnetic fields moving in two directions perpendicular to on another. It is the counterpart of twin-armature rotary-linear motor.

A moving magnetic flux is the main quantity responsible for induction motor performance. In induction motors with two degrees of mechanical freedom this flux is represented by the magnetic field wave, whose motion is described by two space coordinates. Such a description is more general than the one of linear or rotating machines. The rotating magnetic field wave of rotating machines and the traveling field of linear motors are in the mathematical description special cases of the rotating-traveling field of rotary-linear motors. Thus the mathematical model of the motor with two degrees of mechanical freedom can be reduced at any time to the model either of rotary or linear motors.

In this book the mathematical model of the motor with two degrees of mechanical freedom is derived first. Then, introducing the reduction of the mathematical description into the one space coordinate, examples are shown how the linear motor and the rotary motor can be analyzed by means of the more general model of the motor with two degrees of mechanical freedom.

The phenomena, which highly influence the performance of linear motors, in particular those moving with the high speed, are called edge effects. These phenomena, though in a small scale, also occur in rotating machines at the motor edges. Furthermore the edge effects exist in the motor with two degrees of mechanical freedom, but in a more complex way. This is because the moving part of X-Y motor moves with respect to all motor edges. In the book these edge effects are mathematically described, firstly, in a more general way, for machines with two degrees of mechanical freedom, then, by means of the above mentioned reduction, are considered for rotary and linear motors.

The analysis of the motor operation at steady state condition, as well as in dynamics is carried out by using the equivalent circuit. It can also be done by applying the field theory. This, however, requires a very complex tool, in particular, when numerical methods e.g. finite element method is used. When the circuit theory is applied, then, in order to take into account the influence of the edge effects, the parameters of the equivalent circuit are determined from a 3-D field model. To determine the magnetic field in the motor both analytical and numerical methods are used. Numerical methods allow taking into account not only the nonlinearity of the motor iron core but also its geometrical structure (slot dimensions and finite stator and rotor length). Some analytical methods, like that one based on Fourier series technique, allow the consideration of, in an equivalent way [51], the motor geometry. However, the local nonlinearities are hard to include. In practice, this is done by introducing the average magnetic permeability for the whole layer.

There are, however, some advantages of using analytical methods, in particular the Fourier series technique. Suppose the magnetic field is described in a 3-D space by using the latter mentioned technique. In such a case the resultant magnetic field, which is deformed from the sinusoidal by the motor edges, is a sum of a number of harmonics, i.e., sinusoidal waves moving in different directions if the motors with two degrees of mechanical freedom are considered. Such a description of the motor in a field theory allows passing directly to the multi-harmonic equivalent circuit, and analyzing the motor behavior in steady or dynamic conditions with the inclusion of all the effects taking place at the motor edges. Such an approach is presented in this book.

The induction motor described in the field theory is represented by the multilayer structure if the analytical methods are used. In this case the nonuniform geometrical structure of teethed zones of stator and the rotor as well as the air-gap are represented by substituting layers of the calculated, equivalent parameters. The field deformation caused by the teeth and finite dimensions of the stator and the rotors is taken into account by the Fourier technique.

Referring to the above reasoning the text of the book is arranged in the following scheme:
- Description of the induction motor with the magnetic field represented by the rotating-traveling wave,
- Mathematical description of the multilayer and multiharmonic motor in the field theory,
- Equivalent circuit of the multiharmonic motor with the rotating-traveling magnetic field,
- Calculation examples of modeling of particular motors with two- and one- degree of mechanical freedom.

Chapter 2

MONO-HARMONIC INDUCTION MOTOR WITH TWO DEGREES OF MECHANICAL FREEDOM (IM-2DMF)

1.   IDEA OF MOTOR AND MAGNETIC FIELD DESCRIPTION

In IM-2DMF the motion of the magnetic field and the moving part of the motor are described by two space co-ordinates. The magnetic field of an idealized IM-2DMF can be presented as the magnetic flux density wave moving in the direction placed between two co-ordinates as shown in Fig. 2-1. This wave is described by the function

$$B = B_m \exp\left[j\left(\omega t - \frac{\pi}{\tau_x}x - \frac{\pi}{\tau_z}z \right) \right] \tag{2.1}$$

where
   τ_x - pole pitch along x axis,
   τ_z - pole pitch along z axis.

The physical counterpart of the motor with such a magnetic field is the flat X-Y motor with the winding in its primary part twisted with respect to the x and y axes of the coordinate system. The magnetic field interacts with the eddy currents induced in the secondary part, producing the force that acts in the direction of the field motion.

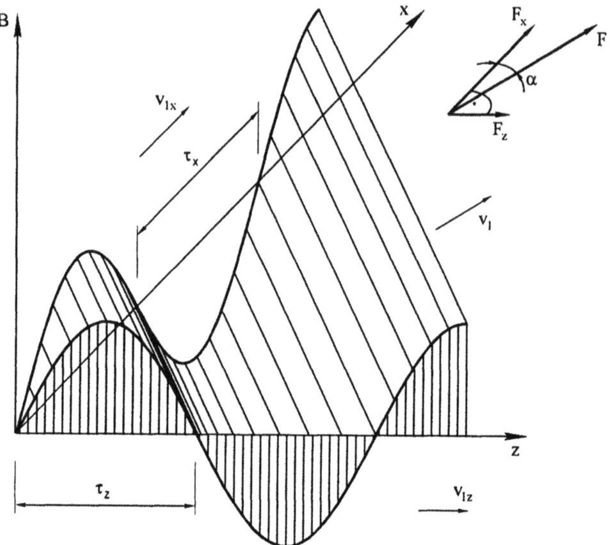

Figure 2-1. Magnetic flux density wave in IM-2DMF

The cylindrical counterpart of such a motor, the rotary-linear motor, is shown in Fig. 2-2. The rotor is a conducting shaft. The stator has a twisted winding, which generates a magnetic flux moving with a helical motion. This rotating-traveling field can be described in the cylindrical co-ordinate system by the function

$$B = B_m \exp\left[j\left(\omega t - \frac{\pi}{\tau_\varphi}\varphi - \frac{\pi}{\tau_z}z \right)\right] \tag{2.2}$$

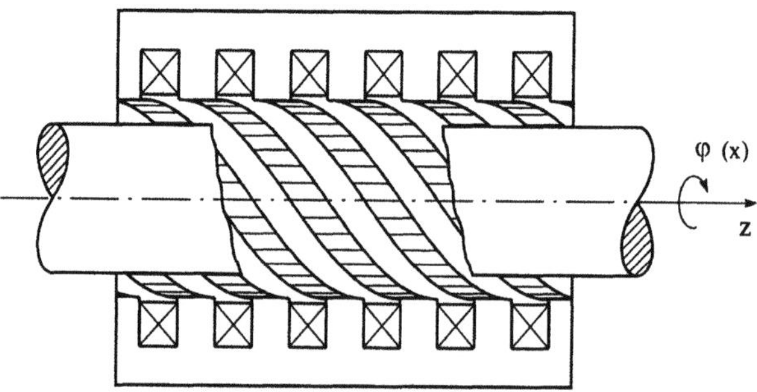

Figure 2-2. Rotary-linear induction motor with rotating-traveling magnetic field

The electromagnetic force F (see Fig. 2-1) acting on the secondary can be divided into two components F_x and F_z. A relation between the forces is as follows:

$$F_x = F \cos \alpha$$
$$F_z = F \sin \alpha \qquad (2.3)$$

Hence the ratio of force components is

$$\frac{F_x}{F_z} = \operatorname{ctg} \alpha \qquad (2.4)$$

Since, according to Fig.2.1,

$$\operatorname{ctg} \alpha = \frac{\tau_z}{\tau_x}, \qquad (2.5)$$

the relation 2.4 takes the form:

$$\frac{F_x}{F_z} = \frac{\tau_z}{\tau_x} \qquad (2.6)$$

In general, the secondary can move in a direction different to the direction of field motion. For example, if the rotor motion is blocked on the direction of the z axis (Fig. 2-2), then it moves only with rotary motion. The direction of rotor motion depends on the loading force. In general, the rotor speed changes during the motor operation not only in its absolute value but in its direction as well. The rotor slip function should take into account these two changes.

2. ROTOR SLIP

To find a formula for the slip let the secondary move with the velocity v (Fig. 2-3) in the direction different from that of the field velocity v_1. Since the secondary speed component v_n, normal to the wave front, decides the induction phenomenon only in the infinitely long motor, the slip s_{xz} can be defined as follows:

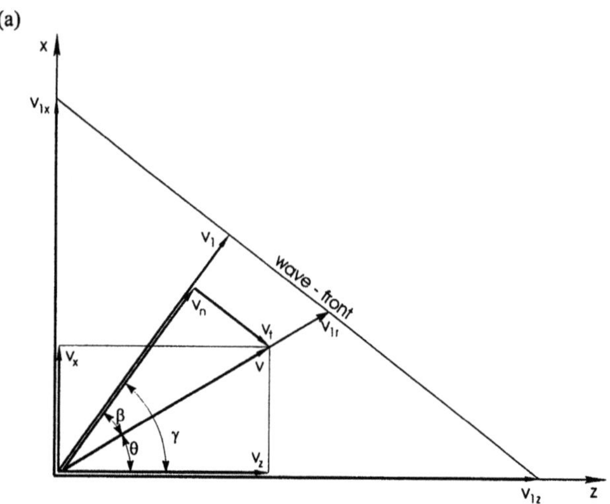

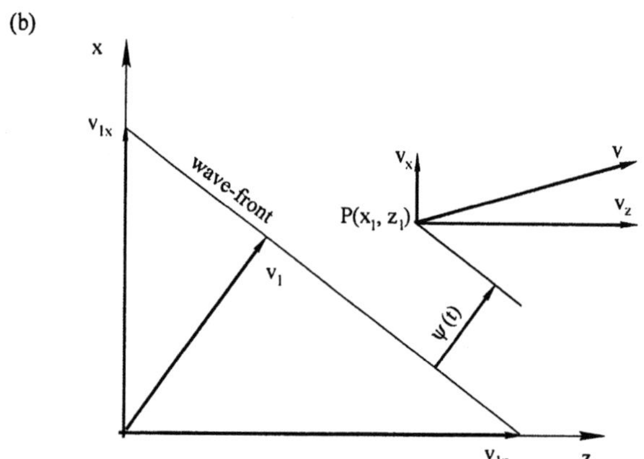

Figure 2-3. Explanation to derivation of rotary-linear slip s_{xz}

$$s_{xz} \overset{df}{=} \frac{v_1 - v_n}{v_1} \tag{2.7}$$

where

$$v_n = v \cos \beta \tag{2.8}$$

Since $\beta = \gamma - \theta$, Eqn. 2.8 takes the form:

$$v_n = v\cos\gamma\cos\theta + v\sin\gamma\sin\theta \qquad (2.9)$$

The secondary speed components are:

$$v_z = v\cos\theta \qquad \text{and} \qquad v_x = v\sin\theta$$

Hence

$$v_n = v_z\cos\gamma + v_x\sin\gamma \qquad (2.10)$$

Besides, from Fig. 2-3

$$v_1 = v_{z1}\cos\gamma \qquad (2.11)$$

Inserting Eqns. 2.10 and 2.11 into 2.7 the slip takes the form

$$s_{xz} = 1 - \frac{v_z + v_x\tan\gamma}{v_{1z}} \qquad (2.12)$$

Since $tg\gamma = v_{z1}/v_{x1}$, Eqn. 2.7 finally takes the form

$$s_{xz} = 1 - \frac{v_x}{v_{1x}} - \frac{v_z}{v_{1z}} \qquad (2.13)$$

where:
　　v_x, v_z – components of rotor speed,
　　v_{1x}, v_{1z} – field velocities along x and z axes expressed by the equations:

$$v_{1x} = 2\tau_x f, \qquad\qquad v_{1z} = 2\tau_z f \qquad (2.14)$$

The slip written in the form above is a function of two rotor speed components v_x and v_z and two adequate speeds of the magnetic field. As was mentioned above, the direction and the modulus of the rotor speed can vary during the motor operation. If, for example, the rotor motion in the z direction is blocked (the axial motion in the rotary-linear motor) the v_z component drops to zero and the slip takes the form:

$$S_{xz} = 1 - \frac{v_x}{v_{1x}}, \tag{2.15}$$

which is well known from the theory of conventional motors.

The Eqn. 2.13 can also be obtained by starting from the following definition of the rotary-linear slip:

$$S_{xz} = 1 - \frac{v}{v_{1r}} \tag{2.16}$$

where v_{1r} is the synchronous speed in the direction of the rotor motion (Fig. 2-3).

There is another way to derive the slip Eqn. 2.13. If the rotor moves with asynchronous speed in relation to the stator field, it means that this field varies for each point on the rotor surface. Thus, for the moving point $P(x_1,z_1)$ on the rotor surface (Fig. 2-3.b):

$$B(t, x_1, z_1) = B_m \exp\left[j\left(\omega t - \frac{\pi}{\tau_x} x_1 - \frac{\pi}{\tau_z} z_1 \right) \right] = variable \tag{2.17}$$

Hence

$$\omega t - \frac{\pi}{\tau_x} x_1 - \frac{\pi}{\tau_z} z_1 = \psi(t) \tag{2.18}$$

where $\psi(t)$ is the angle between point P and the wave-front of the magnetic field wave.

Differentiating each side of equation above, one can get

$$\omega - \frac{\pi}{v_{1x}} v_x - \frac{\pi}{v_{1z}} v_z = \omega_2 \tag{2.19}$$

where v_x, v_z are the rotary and linear speeds, respectively, and

$$\omega_2 = \frac{d\psi(t)}{dt} \tag{2.20}$$

is the angular speed of point P in relation to the stator field.

Similarly, as in the theory of conventional induction motors, it can be written:

$$\omega_2 = \omega s \tag{2.21}$$

Thus, from Eqns. 2.19 and 2.21, the Eqn. 2.13 is obtained.

If the motor work takes place at constant slip then one speed component only depends on the other. For example, if the slip sxz = 0 - for the synchronous version of motor – then from Eqn. 2.8 and 2.9 the following relation is obtained:

$$v_x = \frac{\tau_x}{\tau_z}(v_{1z} - v_z) \tag{2.22}$$

Since τ_x, τ_z and v_{1z} are constants, the two rotor speed components only depend on one another. In the case of the rotary-linear motor, the decreasing of the axial speed (v_z) contributes to the increase of rotary speed (v_x) and vice versa. The transformation of one speed into another takes place.

3.   ELECTROMAGNETIC FIELD IN A MULTILAYER MOTOR

To determine the performance characteristics the most common approach is to calculate them from the equivalent circuit of the motor. The parameters of the circuit need to be determined from the electromagnetic field, which is found from the field model of the motor. As was written in the Introduction, to find the field components, the analytical method is used that is based on Fourier's series technique. In this method the motor is represented by the N layer structure shown in Fig. 2-4.

Each layer is homogeneous, isotropic and linear and defined by magnetic permeability μ_i, conductivity γ_i and velocity v_i. The primary winding is reduced to an infinitely thin current sheet placed between the second and the third layer. The first layer corresponds to the teeth zone with a saturation represented by the finite value of permeability μ_2. The current density of the current layer is defined by the function:

$$J^s = J^s_m \exp[j(\omega t - \frac{\pi}{\tau_x}x - \frac{\pi}{\tau_z}z)] \tag{2.23}$$

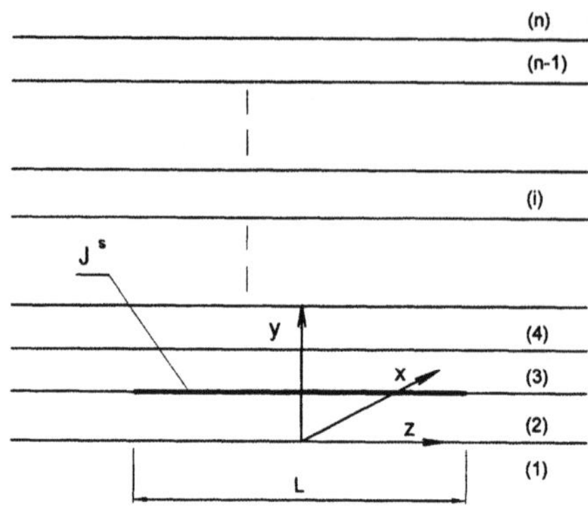

Figure 2-4. N-layer motor model

where J_m^s is the amplitude of linear current density and $\omega=2\pi f$, is the angular frequency.

In the homogeneous space the electromagnetic field is determined from the Maxwell's equations written in the form:

$$\nabla \times \vec{H} = \vec{J} \tag{2.24}$$

$$\nabla \times \vec{E} = -\frac{\partial \vec{B}}{\partial t} \tag{2.25}$$

$$\nabla \cdot \vec{B} = 0, \; \nabla \cdot \vec{D} = 0 \tag{2.26}$$

For the isotropic and homogeneous environment:

$$\vec{D} = \varepsilon \vec{E}, \; \vec{B} = \mu \vec{H} \tag{2.27}$$

Besides, for the moving area

$$\vec{J} = \gamma(\vec{E} + \vec{v} \times \vec{B}) \tag{2.28}$$

When the co-ordinate system is attached to the moving area then frequency $\omega_{s_{xz}} = 2\pi s_{xz} f$ is expressed in terms of the slip s_{xz}, $B = B(\omega s_{xz}, t)$ and the Eqn. 2.28 can be reduced to $\vec{J} = \gamma \vec{E}$ [62].

For convenience the vector potential A is used. In the problem discussed here two A components occur: A_x and A_z. The transformed Maxwell's equations for the *i-th* layer expressed in the complex number form are as follows:

$$\nabla^2 A_{mxi} = \alpha_i^2 A_{mxi}$$
$$\nabla^2 A_{mzi} = \alpha_i^2 A_{mzi}$$

(2.29)

where:

$$\alpha_i^2 = j\omega s_{ixz}\gamma_i\mu_i$$

(2.30)

The relations between the electric field intensity E and the magnetic flux density B and the vector potential A in the *i-th* layer are expressed by the equations:

$$B_{mi} = \Delta \times A_{mi}$$

$$E_{mi} = -j\omega A_{mi}$$

(2.31)

Since there are no y-directed currents in the computational model, the components E_y and A_y are assumed to be zero. The final solutions of Eqn. 2.29, derived for i-th layer are as follows (see Appendix II):

$$E_{mxi} = K_i'(y)F(x,z)$$

$$E_{mzi} = -\frac{\tau_z}{\tau_x}K_i'(y)F(x,z)$$

$$B_{mxi} = -j\frac{\beta_i}{\omega}\frac{\tau_z}{\tau_x}K_i''(y)F(x,z)$$

(2.32)

$$B_{myi} = \frac{\alpha_o}{\omega}\frac{\tau_z}{\pi}K_i'(y)F(x,z)$$

$$B_{mzi} = -j\frac{\beta_i}{\omega}K_i''(y)F(x,z)$$

where K_i' and K_i'' are coefficients that are functions of the y component and design parameters. Besides:

$$F(x,z) = \exp[-j(\frac{\pi}{\tau_x}x + \frac{\pi}{\tau_z}z)] \tag{2.33}$$

The electromagnetic field components were derived in the coordinate system attached to the secondary layers. Thus the current density components in the secondary *i*-th layer are calculated in the following way:

$$J_{mix(z)} = \gamma_i E_{ix(z)} \tag{2.34}$$

4. MOTOR EQUIVALENT CIRCUIT

An assumption of constant current density in the field model is synonymous to the steady current source that supplies the motor. This approach is commonly used in the field analysis of linear motors when the coils of each phase are connected in series. With such a connection the current density distribution along the motor is independent on the speed, while the magnetic field distribution varies. This field variation is generated by the edge effects.

In conventional rotary motors, when the symmetries of magnetic and electric circuits are assured the magnetic field distribution in the air-gap is constant. The field analysis can be carried out starting with the assumption of the known magnetic flux density in the air-gap. This is synonymous to the steady voltage source that supplies the motor winding.

When a steady current source is applied, the magnetic flux density in the air-gap changes with the speed, what contributes to the variation of the voltage across the winding terminals. If the motor is fed from the steady voltage source the speed variation causes the change of primary and secondary currents.

In the infinitely long and wide flat IM-2DMF the magnetic field distribution in the air-gap is constant and the field analysis could be carried out under the assumption of constant magnetomotive force in the air-gap. Since the edge effects are taken into account (in the next section), the electromagnetic field components will be derived assuming the known primary current density.

There are differences in analysis of the motor performance if the motor is supplied from a voltage or current source. Due to that the analysis of the motor performance at steady state conditions will be carried out separately for the motor fed from a current and a voltage source.

4.1   Motor fed from a steady current source

To determine the motor performance at constant primary current I_1 it is sufficient to apply the equivalent circuit without the parameters R_1 and X_{s1} of the primary winding, since they do not influence the motor performance characteristics. Thus the equivalent circuit that can be used does not contain the stator winding parameters and it looks like the circuit shown in Fig. 2-5.

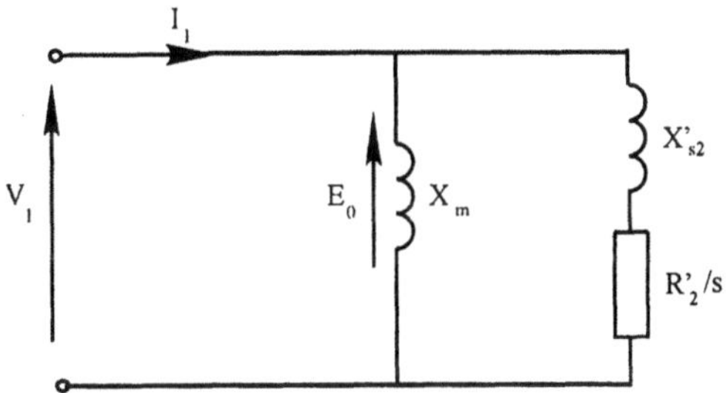

Figure 2-5. Equivalent circuit of IM-2DMF supplied from the current source

Since the relations that describe the electromagnetic field components are the functions of the primary current (contained in the primary current density expressions), the power crossing the air-gap, the power losses, as well as the forces acting on the secondary layers can be determined directly from the electromagnetic field equations without necessity of passing to the circuit theory. Due to that, most of the authors of publications dealing with this type of problem calculate the performance characteristics at steady current supply without using the circuit model [1, 58, 63].

In the motor analyzed here, three magnetic field components and two electric field components appear. That contributes to the rise of three force components. The average in time values of force densities can be calculated from the equations [61]:

$$f_x = 0.5\gamma \operatorname{Re}\{-E_{mz} B_{my}^*\}$$
$$f_y = 0.5\gamma \operatorname{Re}\{E_{mz} B_{mx}^* - E_{mx} B_{mz}^*\} \tag{2.35}$$
$$f_z = 0.5\gamma \operatorname{Re}\{-E_{mx} B_{my}^*\}$$

The component f_y is the force normal to the secondary surface (levitation force in X-Y motor), and f_x and f_z are tangential components that drive the motor. The entire values of these forces are being found by integration of formulae 2.35 over the whole volume of the secondary. The driving forces F_x and F_z may also be determined from the expressions applied in the theory of induction motors. In that case the power crossing the air-gap P_r, the magnetic field speed v_1 and the force F_{xz} acting on the secondary are linked by the relation:

$$P_r = v_1 F_{xz} \tag{2.36}$$

Since the power

$$P_r = \frac{\Delta P_2}{s_{xz}}, \tag{2.37}$$

hence the force

$$F_{xz} = \frac{\Delta P_2}{v_1 s_{xz}} \tag{2.38}$$

where ΔP_2 are the secondary power losses.

The power losses can be derived applying Poynting's vector $\vec{S}$ [61]. Its normal component S_n on the secondary surface is as follows

$$S_n = 0.5 \left[E_{mz3} H_{mx3}^* - E_{mx3} H_{mz3}^* \right]_{y=g} \tag{2.39}$$

When the coordinate system is attached to the primary, S_n is equal to the power entering the secondary (i.e. mechanical power + secondary power losses). Here, the coordinates were linked with the secondary layers and the Poynting's vector component is equal to the secondary power loss density. The entire power losses are equal to the real part of the integral calculated over the secondary surface Q according to the equation:

$$\Delta P = \mathrm{Re} \left\{ \iint_Q S_n \, dq \right\} \tag{2.40}$$

Referring to the IM-2DMF the Eqn. 2.38 is transformed into two expressions related to two force components:

$$F_x = \frac{\Delta P_2}{v_{1x} s_{xz}}, \qquad F_z = \frac{\Delta P_2}{v_{1z} s_{xz}} \tag{2.41}$$

The electromotive force E_o (see Fig. 2-5) induced in the primary winding by the magnetic flux Φ_o passing the current layer can be determine from the expression:

$$E_o = 4.44\, fT^{(p)} K_w \Phi_o \tag{2.42}$$

where:

$$\Phi_o = \int_0^{\tau_x} \int_0^{\tau_z} B_{my2}(x,y,z)_{y=0}\, dxdz \tag{2.43}$$

    – is the magnetic flux at synchronous speed ($s_{xz} = 0$),

B_{my2} – is the normal component of magnetic flux density on the current layer surface,

K_w – is the winding coefficient,

$T^{(p)}$ – number of turns per phase.

In order to find the voltage across the primary terminals it is necessary to determine the primary resistance and leakage reactance. These parameters can be found using the methods applied in conventional motors. The phase voltage across the winding terminals is equal to (see Fig. 2-6):

$$V_1 = E_0 + I_1(R_1 + jX_{s1}) \tag{2.44}$$

where E_o is changing with the slip at constant current.

4.2 Motor supplied from a steady voltage source

An analysis of the performance of a motor supplied from a steady voltage source is carried out most often by means of the equivalent circuit. When the secondary source consists of homogeneous layers, as in the motor considered here, the circuit parameters are determined from the electromagnetic field components.

The secondary circuit can be represented by the network of impedances of current paths. This method applied sometimes to rotary motors with solid rotors [5, 8], can be used in the case of uniformly distributed currents in the secondary. Due to the edge effects, the current distribution in the secondary is not uniform. In that case the secondary parameters are calculated from the active and reactive power losses [62]. This approach is applied here.

The equivalent circuit of the IM-2DMF shown in Fig. 2-6 corresponds with the one well known from the theory of rotary induction motors. The parameters shown there are:

R_1, X_{s1} – resistance and leakage reactance of the primary,
$X_{m.}$ – magnetising reactance,
R_{Fe} – resistance, representing the primary core losses,
R'_2, X'_2 – resistance and leakage reactance of the secondary referred to the primary side.

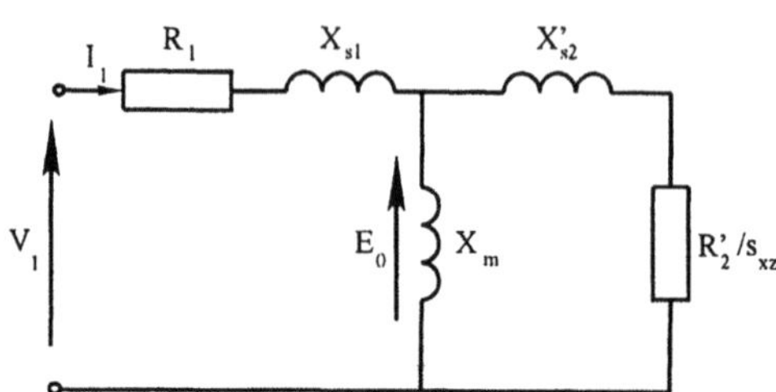

Figure 2-6. Equivalent circuit of IM-2DMF supplied from voltage source

The secondary parameters can be determined from the equations:

$$R_2 = \frac{\Delta P_2}{m|I_2|^2}$$
(2.45)

$$X_{s2} = \frac{\Delta Q_2}{m|I_2|^2}$$
(2.46)

where:

ΔP_2 and ΔQ_2 – active and reactive power losses in the secondary,
m – number of primary phases,
I_2 – secondary equivalent current related to the primary side.

The secondary active and reactive power losses can be determined from Poynting's vector defined by the Eqn. 2.39 and quoted again here:

$$S_n = 0.5 \left[E_{mz3} H^*_{mx3} - E_{mx3} H^*_{mz3} \right]_{y=g}$$

It represents the apparent power crossing the secondary surface unit. Since the coordinate system of the field model in section 2 was attached to the secondary, this apparent power does not contain the mechanical power. It is equal to the surface density of the secondary power losses. Thus, the entire active and reactive power losses are calculated as the integrals over the secondary surface A:

$$\Delta P^r = \iint_A \mathrm{Re}\{S_n\}\, da \tag{2.47}$$

$$\Delta Q^r = \iint_A \mathrm{Im}\{S_n\}\, da \tag{2.48}$$

The equivalent secondary current referred to the primary winding can be determined from the equation:

$$I_2' = \frac{F_m^r}{F_m^s} I_1 \tag{2.49}$$

where:

I_1 – phase primary current,

F_m^s, F_m^r – amplitudes of primary and secondary magnetomotive forces. For the nth motor layer these are calculated from the equations:

$$F_m^s = \int_0^x J_{mz}(x,z)\, dx \tag{2.50}$$

$$\begin{aligned}
F_m^r = &\int_0^x \int_g^{g+d_4} \gamma_4 E_{mz4}(x,z)\, dx\, dy + \int_0^x \int_{g+d_4}^{g+d_4+d_5} \gamma_5 E_{mz5}(x,z)\, dx\, dy + \ldots + \\[2mm]
&+ \int_0^x \int_{g+d_4+\ldots+d_{n-1}}^{g+d_4+\ldots+d_{n-1}+d_n} \gamma_n\, E_{mzn}(x,z)\, dx\, dy
\end{aligned} \tag{2.51}$$

The magnetizing reactance X_m is calculated as the ratio:

$$X_m = \frac{E_o}{I_m} \tag{2.52}$$

where E_o is the electromotive force expressed by Eqn. 2.42 and the magnetizing current I_m is found from the equation:

$$I_m = \frac{F_m^m}{F_m^s} I_1 \qquad\qquad (2.53)$$

where:

$$F_m^m = F_m^s + F_m^r \qquad\qquad (2.54)$$

The equivalent circuit shown in Fig. 2-6 has the secondary branch common for all the secondary layers. It can be split into separate parallel branches representing adequate layers. In the case of a double layer secondary, consisting of e.g. copper and back iron, the secondary circuit is similar to the circuit of a double-cage rotor as shown in Fig. 2-7.

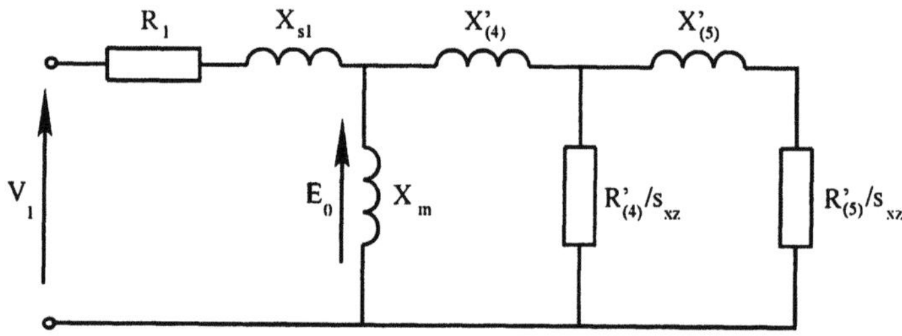

Figure 2-7. Equivalent circuit of IM-2DMF with double-layer secondary

In this case the secondary parameters are calculated from the equations:

$$R_{(4)}' = \frac{\Delta P_{(4)}}{m\left|I_{(4)}'\right|^2}, \qquad X_{(4)}' = \frac{\Delta Q_{(4)}}{m\left|I_{(4,5)}'\right|^2} \qquad\qquad (2.55)$$

The active and reactive power losses in the two secondary layers are:

$$\Delta P_{(5)} = \iint_{Q} \mathrm{Re}\left\{S_{(5)}\right\} dq, \qquad \Delta Q_{(5)} = \iint_{Q} \mathrm{Im}\left\{S_{(5)}\right\} dq \qquad\qquad (2.56)$$

$$\Delta P_{(4)} = \Delta P^r - \Delta P_{(5)}, \qquad \Delta Q_{(4)} = \Delta Q^r - \Delta Q_{(5)} \qquad\qquad (2.57)$$

The entire power losses in the secondary ΔP^r and ΔQ^r are expressed by the Eqns. 2.47 and 2.48. The surface density of the apparent power crossing the 5th layer is

$$S_{(5)} = 0.5\left[E_{mz5}H_{mx5}^{*} - E_{mx5}H_{mz5}^{*}\right]_{y=g+d_4} \tag{2.58}$$

The current of the two secondary layers $I_{(4,5)}$ and the current of the 5th layer can be calculated as follows:

$$I'_{(4,5)} = \frac{F_{(4,5)}}{F_m^s}I_1, \quad I'_{(5)} = \frac{F_{(5)}}{F_m^s}I_1 \tag{2.59}$$

where the magnetomotive forces are

$$F_{(4,5)} = \int_0^x \int_g^{g+d_4} \gamma_4 E_{mz4}(x,z)dxdy + \int_0^x \int_{g+d_4}^{g+d_4+d_5} \gamma_5 E_{mz5}(x,z)dxdy,$$

$$F_{(5)} = \int_0^x \int_{g+d_4}^{g+d_4+d_5} \gamma_5 E_{mz5}(x,z)dxdy \tag{2.60}$$

Similarly as for rotary motors, the secondary resistance R_2'/s_{xz} of the equivalent circuit shown in Fig. 2-6 can be split into two resistances as shown in Fig. 2-8.

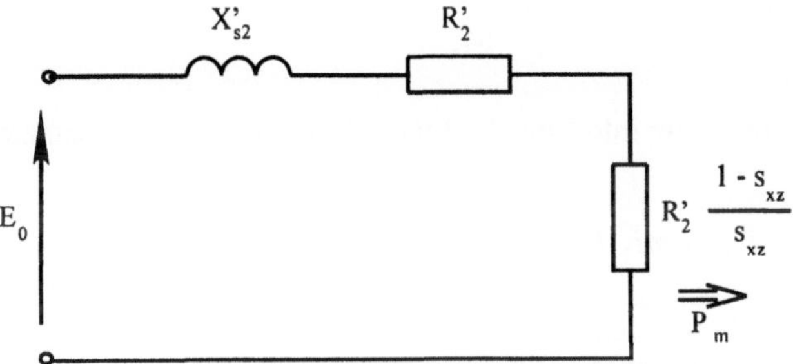

Figure 2-8. Equivalent circuit of rotor of IM-2DMF

Since the motor moves in the direction between x and z axes the resistance $R_2'\dfrac{1-s_{xz}}{s_{xz}}$ represents the mechanical power of the resultant motion. This resistance can be divided into two resistances representing the mechanical power components of the two motion components along x and z axes in the following way:

The entire mechanical power P_m is a sum of two components:

$$P_m = P_{mx} + P_{mz}$$

(2.61)

where

P_{mx} – the mechanical power of motion component in x direction, and

P_{mz} – the mechanical power of motion component in z direction.

From the circuit shown in Fig. 2-8 the mechanical power is:

$$P_m = mR_2' \frac{1 - s_{xz}}{s_{xz}} I_2'^2$$

(2.62)

where m is the number of phases.

Inserting Eqn. 2.13 into 2.62 the following expression is obtained:

$$P_m = mI_2'^2 \frac{R_2'}{s_{xz}} \left(\frac{v_x}{v_{1x}} + \frac{v_z}{v_{1z}} \right)$$

(2.63)

Defining the following quantities:

$$s_x = 1 - \frac{v_x}{v_{1x}} \text{ and } s_z = 1 - \frac{v_z}{v_{1z}}$$

(2.64)

and inserting them into Eqn. 2.63 the following relation is obtained:

$$P_{mx} + P_{mz} = mI_2'^2 R_2' \frac{1 - s_x}{s_{xz}} + mI_2'^2 R_2' \frac{1 - s_z}{s_{xz}}$$

(2.65)

After a transformation the new secondary circuit is as shown in Fig. 2-9. In the case of the rotary-linear motor (see Fig. 2-2) the P_{mx} is the mechanical power of rotary motion and P_{mz} is the mechanical power of linear motion. Fig. 2-10 is a power flow diagram illustrating the mechanical power division.

The parameters of the secondary circuit are functions of the slip. To determine the electromechanical characteristics on the basis of the equivalent circuit it is necessary to determine their values for the entire range of the speed that is concerned. In the case of the double-layer secondary (copper and iron) the resistance of a conducting layer, at its certain thickness, has the predominant influence on motor performance. Since this

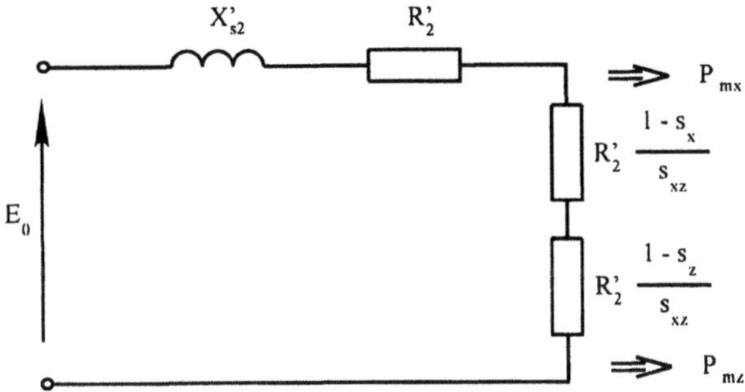

Figure 2-9. Equivalent circuit of rotor of IM-2DMF with "mechanical" resistances split into two components

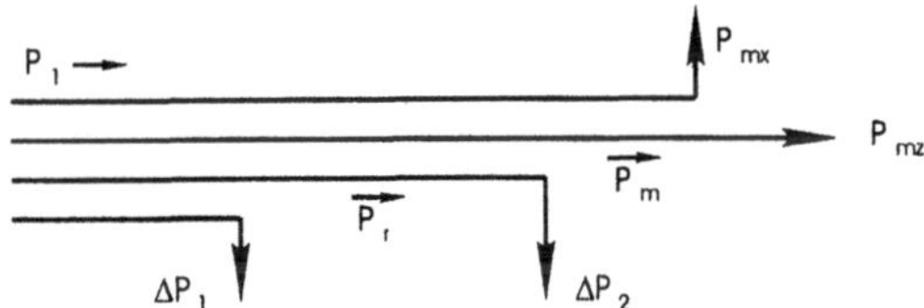

Figure 2-10. Power flow diagram of IM-2DMF

resistance is nearly constant, the electromechanical characteristics calculated at constant parameters do not differ much from those found when they vary with speed.

5.　　PERFORMANCE CHARACTERISTICS

The characteristics of all electromechanical quantities when are drawn as functions of slip s_{xz} or speed components v_x and v_z have the shape of a surface, not curves as for IM-1DMF. As an example, a force-slip characteristic is shown in Fig. 2-11.

Considering the steady state operation of the machine set: motor + load, the operating point depends on the mechanical characteristics of both machines. In the case of the IM-2DMF the work of such a machine set depends on its mechanical characteristics, which are the functions of speed components v_x and v_z.

In order to determine the operating point of the machine set, let be considered the rotary-linear motor with the rotating-traveling field. Let the rotor be loaded by two machines acting independently on linear (axial) and rotational directions with the load force characteristics shown in Fig. 2-12.

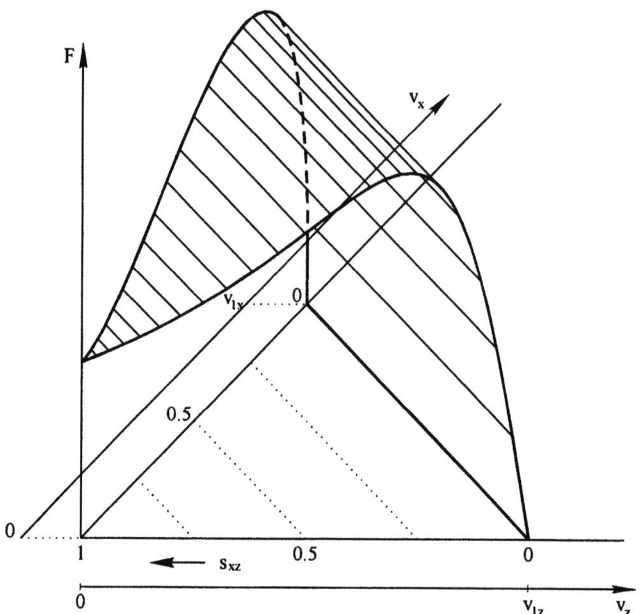

Figure 2-11. Force-slip characteristic of IM-2DMF

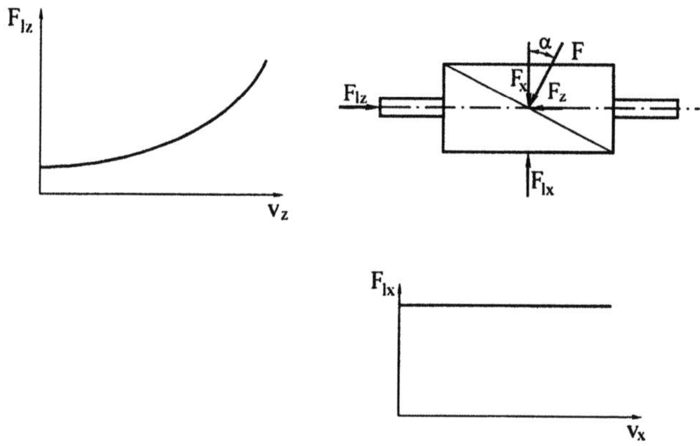

Figure 2-12. Load characteristics for IM-2DMF

The equilibrium of the machine set takes place when the resultant load force is equal in its absolute value and opposite to the force developed by the motor. The direction of the electromagnetic force of the motor is constant according to the relation 2.6 and does not depend on the load. Thus, at steady state operation both load forces F_{lx} and F_{lz} acts against motor force components F_x and F_z in the same direction if the following relation between them takes place:

$$\frac{F_x}{F_z} = \frac{F_{lx}}{F_{lz}} = \text{ctg}\,\alpha = \frac{\tau_z}{\tau_x}. \tag{2.66}$$

In order to determine graphically the operating point where the above relation holds, it is necessary to draw all mechanical characteristics on a common graph as functions of speed components v_x and v_z, as shown in Fig. 2-13. To compare and to draw both load characteristics on a common graph, the real load forces F_{lx} and F_{lz} acting separately on rotational and linear directions should be transformed into F_{lx}' and F_{lz}', forces acting on the direction of the motor force F. These equivalent forces are

$$F_{lx}' = \frac{F_{lx}}{\cos\alpha} \text{ and } F_{lz}' = \frac{F_{lz}}{\sin\alpha}. \tag{2.67}$$

The transformed load characteristics drawn as a function of v_x and v_z, as shown in Fig. 2-13, are the surfaces intersecting one another along the section $\overline{KA}$. This section that forms the $F_l(v)$ characteristic is a set of points where the following equation is fulfilled:

$$F_{lx}' = F_{lz}' \tag{2.68}$$

Inserting Eqns. 2.67 into 2.68 the relation 2.66 is obtained.

The load characteristic $F_l(v)$ intersects with the motor characteristic at points A and B, where the equilibrium of the whole machine set takes place. To check if the two points are stable the steady state stability criterion can be used, which is applied to rotary motors in the following form:

$$\frac{dF}{dv} < \frac{dF_l(v_x, v_z)}{dv} \tag{2.69}$$

at the assumption

$$\frac{dF_{lx}}{dv_x} > 0 \text{ and } \frac{dF_{lz}}{dv_z} > 0 \tag{2.70}$$

Applying this criterion, point A in Fig. 2-13 is a stable one and point B is unstable. More on that can be found in [43].

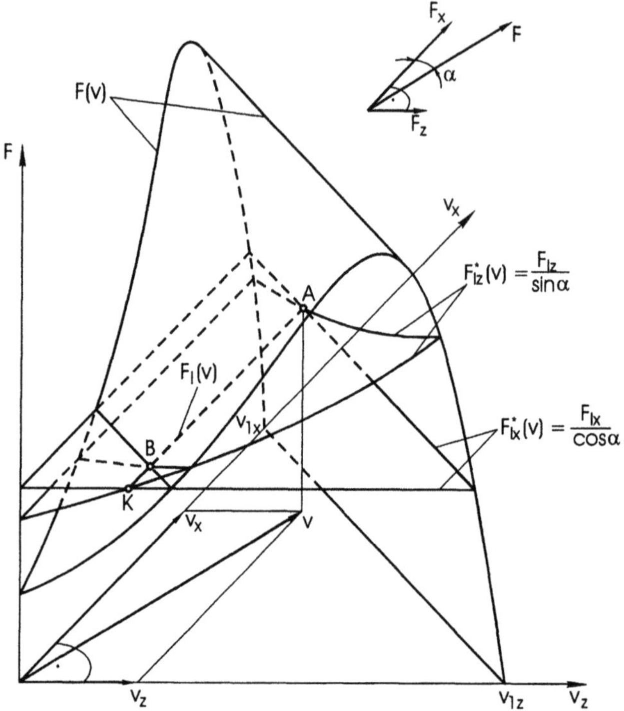

Figure 2-13. Determination of the operating point A of the machine set with IM-2DMF

6.   CONVERSION OF MATHEMATICAL MODEL OF IM-2DMF INTO ONE OF IM-1DMF

Referring to Fig. 2-1 one can see that the change of motion direction of the field wave contributes to the change of pole pitches τ_x and τ_z if the wave length remains steady. In the case of the turning of the wave towards the z direction the pole pitch τ_z is decreasing, while the pitch τ_x is increasing. When the wave motion goes on along the z direction then the pole pitch τ_z equals half of the wave-length and the pitch τ_x reaches infinity. Inserting into function 2.1 $\tau_x=\infty$, the field description expressed by two space coordinates turns to the description of the magnetic field of IM-1DMF, referred to as a flat linear motor:

$$B = B_m \exp\left[j\left(\omega t - \frac{\pi}{\tau_z} z \right) \right] \tag{2.71}$$

In the case of the magnetic wave description of a rotary-linear motor by Eqn. 2.2, if $\tau_z = \infty$ is substituted, function $B(x, z)$ changes to the form:

$$B = B_m \exp\left[j\left(\omega t - \frac{\pi}{\tau_\varphi}\varphi \right) \right],\tag{2.72}$$

which describes the magnetic flux density wave of a rotary motor.

Applying the same procedure to the speed relations 2.14, i.e. inserting $\tau_x = \infty$ the speed $v_{1x} = \infty$. Doing the same to the slip equation 2.13, i.e. inserting $v_{1x} = \infty$, the slip turns to the form

$$s_{xz} = 1 - \frac{v_z}{v_{1z}}\tag{2.73}$$

which is well known from the theory of linear or rotary motors.

The force component $F_x = F_z \dfrac{\tau_z}{\tau_x}$ from Eqn. 2.6 equals to zero after substitution $\tau_x = \infty$. Similarly, all the relations derived for IM-2DMF change to the equations valid for IM-1DMF if either $\tau_x = \infty$ or $\tau_z = \infty$ is substituted.

The conclusion from the above reasoning is, that the mathematical model of the IM-2DMF is more general with respect to the model of IM-1DMF and the equations derived for IM-2DMF can be converted to the ones of rotary or linear motors.

Chapter 3

COMPUTATIONAL MODEL OF IM-2DMF TAKING INTO ACCOUNT FINITE DIMENSIONS OF PRIMARY STRUCTURE, AND CURRENT AND WINDING ASYMMETRIES

So far an idealized IM-2DMF with sinusoidal magnetic field in an air-gap and infinitely long and wide, i.e. without any edge effects, has been analyzed. The real motor, in general terms, can be equipped with an asymmetrical winding (e.g. single-phase motor with auxiliary winding) and asymmetrical current flowing in it. Furthermore, the winding is distributed in the slots and the primary or the secondary has finite dimensions. This latter contributes to the non-uniform distribution of the magnetic field in the air-gap called edge effects. A more complicated problem is, if the supply current is nonsinusoidal.

Proposed in this chapter is the mathematical model of the motor taking into account the primary winding and current asymmetry simultaneously with the discrete winding distribution in the slots and the finite width and length of the motor. A more general model is discussed in [18, 19] where additionally the nonsinusoidal current and nonuniform distribution of molten metal in an induction pump is considered.

The computational model should take into account the motor geometry. Among the IM-2DMFs the most complex phenomena occur in the flat X-Y motor and in the motor with the spherical rotor. To find the electromagnetic field, a 3-D field model has to be considered. The rotary-linear motors, due to the circumferential electric and magnetic symmetry can be described using 2-D model, which is contained in 3-D one as its particular case.

1.   3-D FIELD MODEL OF IM-2DMF

The representative of the IM-2DMF is a flat X-Y motor of finite width and length of primary as shown in Fig. 3-1. This motor is represented by the multilayer motor structure with the electric and magnetic properties similar to those, which are in the model defined in chapter 2 section 3. The primary core is infinitely long and wide in the computational model, while the infinitely thin current sheet has the dimensions adequate to the surface of the real armature.

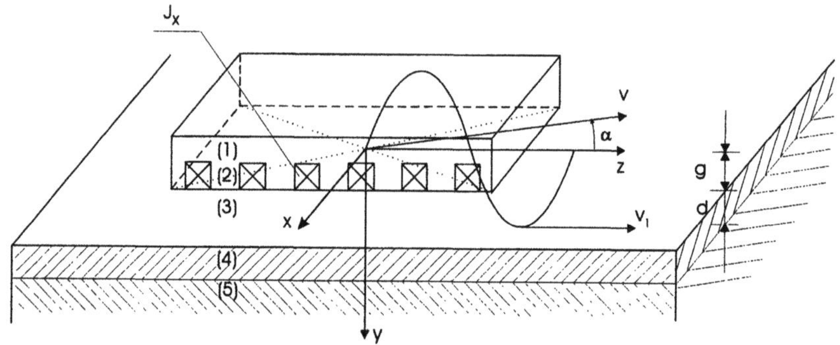

Figure 3-1. Model of IM-2DMF with finite dimensions of primary and secondary parts

There are papers [50, 51] applying the Fourier's series method, which takes into account the finite dimensions of the primary core. The results contained in [51] indicate that this has little effect on motor performance.

In general, the primary current sheet placed between the primary and air-gap consists of two layers representing two windings, which are perpendicular to one another, as it is in one of the constructions of flat X-Y motors shown in Fig. 1-1. One of the current layers has x directed current J_x, the another, symbolized by J_z, contains the z directed current. The influence of end connections of the winding is ignored. The secondary moves with speed v in the direction that forms, with the motion direction of magnetic field, angle α (Fig. 3-1).

It is assumed that each winding consists of m phases supplied by unsymmetrical phase currents. Further consideration will concern the motor with the current layer of the x directed winding only, represented by the current linear density J_x. A similar approach could be taken for current density J_z. The entire electromagnetic field in the motor will be an effect of two windings.

The current density with the discrete winding distribution in the slots is described by the following formula (see Appendix I):

$$J_x(t,z) = \sum_v \sum_h \sum_q \sum_t J_{mxvhqt} \exp\left[j\left(\omega t - \frac{\pi}{\tau_{zvh}} z \right) \right] \tag{3.1}$$

where:

J_{mxvhqt} – amplitude of $hvqt^{th}$ harmonic,

$$\tau_{zvh} = \frac{1}{q\dfrac{2h}{\tau_t} + t\dfrac{v}{\tau_z}}, \qquad q, t = 1, -1, \tag{3.2}$$

τ_t – tooth pitch,
τ_z – pole pitch,
v – 1, 3, 5, 7, ... – phase harmonics,
h – 0, 1, 2, 3, ... – tooth harmonics.

The stator slotting of both windings is included in the motor model by introducing an equivalent air gap g_r defined as

$$g_r = g K_{cx} K_{cz} \tag{3.3}$$

where K_{cx} and K_{cz} are Carter's coefficients for x and z directed slots expressed in the form:

$$K_{cx(z)} = \frac{\tau_{tx(z)}(5g + b_{1x(z)})}{\tau_{tx(z)}(5g + b_{1x(z)}) - b_{1x(z)}^2} \tag{3.4}$$

where: b_1 – slot opening, g – the real air-gap, τ_t – tooth pitch.

As was written above, the finite dimensions of the primary are represented by the finite length L and width W of the current layer. Since the Fourier's method is used to solve differential field equations the current layer in the computational model is represented by a double series of identical layers as shown in Fig. 3-2. This is analytically expressed by the double-series of current density in the following way:

$$J_x(t,z) = \sum_l \sum_k \sum_v \sum_h \sum_q \sum_t J_{mxklvhqt} \exp\left[j\left(\omega t - \frac{\pi}{\tau_{zvh}} z \right) \right] \cdot \cos\left(k\frac{\pi}{\tau_{sx}} x \right) \cos\left(l\frac{\pi}{\tau_{sz}} z \right) \tag{3.5}$$

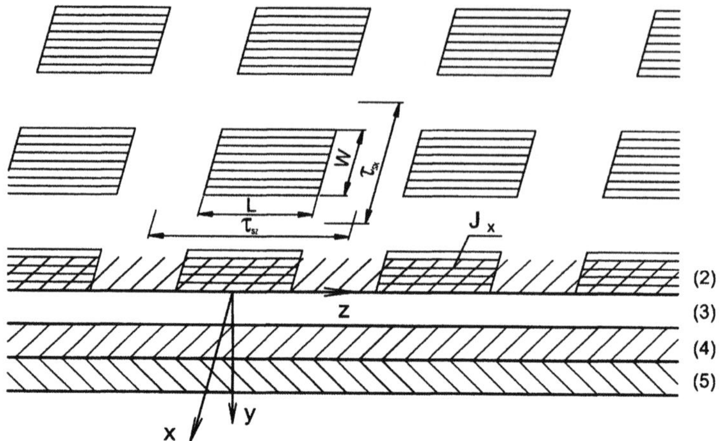

Figure 3-2. Multi-layer field model of IM-2DMF with finite length L and width W of primary current layer

where

$$J_{mxklvhqt} = \frac{16}{\pi^2 kl}\sin\left(k\frac{W\pi}{2\tau_{sx}}\right)\sin\left(l\frac{L\pi}{2\tau_{sz}}\right)J_{mxvhqt} \qquad (3.6)$$

$k, l = 1, 3, 5, \dots$ – harmonics of the double-series of current density
After trigonometric transformation Eqn. 3.5 takes the form

$$J_x(t,z) = \sum_l \sum_k \sum_u \sum_r \sum_v \sum_h \sum_q \sum_t \tfrac{1}{4} J_{mxklvhqt}$$
$$\cdot \exp\left[j\left(\omega t - \frac{\pi}{\tau_{xku}}x - \frac{\pi}{\tau_{zlrvh}}z\right)\right] \qquad (3.7)$$

where

$$\tau_{xku} = \frac{\tau_{sx}}{uk}, \quad \tau_{zhvlr} = \frac{\tau_{zhv}\tau_{sz}}{lr\tau_{zhv}+\tau_{sz}} \qquad (3.8)$$

$u, r = 1, -1$

The symbols τ_{sx} and τ_{sz} are the pitches of fictitious multi-primary model (Fig. 3-2). The differences: $c_x = \tau_{sx} - L$ and $c_z = \tau_{sz} - W$ are the gaps between fictitious primaries. These gaps should have values such that the magnetic field of the preceding primary does not overlap the field of the next ones. The assessment method of these gaps is given in [37].

The current density written in the form 3.7 is a sum of k,l harmonics of J_{xhvqt}, which each is the sum of four waves ($u, r = 1,-1$) of the length $2\tau_{xk}$ and $2\tau_{zhvl}$, moving in the direction defined by two space co-ordinates x and z. The slip of the secondary related to these waves is described by the equation

$$S_{xzklurvhqt} = 1 - \frac{v_x}{v_{xku}} - \frac{v_z}{v_{zlvhr}} \tag{3.9}$$

in which

$$v_{xku} = 2\tau_{xku}f \quad \text{and} \quad v_{zlvhr} = 2\tau_{zlvhr}f \tag{3.10}$$

The electromagnetic field of the motor whose discrete winding distribution and finite length and width are taken into account is calculated as the harmonic sum:

$$E_m = \sum_k \sum_l \sum_u \sum_r \sum_v \sum_h \sum_q \sum_t E_{mklurvhqt} \, ,$$

$$B_m = \sum_k \sum_l \sum_u \sum_r \sum_v \sum_h \sum_q \sum_t B_{mklurvhqt} \tag{3.11}$$

where the $klurvhqt^{th}$ harmonic of the H and B components are calculated from Eqns. 2.32 after substitutions $\tau_x = \tau_{xku}$, $\tau_z = \tau_{zlvhr}$, $S_{xz} = S_{xzklurvhqt}$, $J_{mx} = 0.25 \cdot J_{mxklurvhqt}$.

To calculate the average values of force and power loss densities in the ith secondary layer the following equations are used:

$$\vec{f}_i = \frac{1}{2}\gamma_i \, \mathrm{Re}\left\{\vec{E}_{mi} \times \vec{B}_{mi}^{*}\right\} \tag{3.12}$$

$$\Delta P_i = \frac{1}{2}\gamma_i \, \mathrm{Re}\left\{\vec{E}_{mi} \cdot \vec{E}_{mi}^{*}\right\} \tag{3.13}$$

The total driving forces and secondary power losses can be calculated also as the sums of forces and power losses generated by the field harmonics. Thus:

$$\Delta P = \sum_k \sum_l \sum_u \sum_r \sum_v \sum_h \sum_q \sum_t \Delta P_{klur\,vhqt} \qquad (3.14)$$

$$F_{(n)} = \sum_k \sum_l \sum_u \sum_r \sum_v \sum_h \sum_q \sum_t F_{(n)klur\,vhqt} \qquad n = x,\,z \qquad (3.15)$$

where:

$$F_{(n)klur\,vhqt} = \frac{\Delta P_{klur\,vhqt}}{v_{(n)klur\,vhqt}\,S_{xzklur\,vhqt}} \qquad (3.16)$$

$$\Delta P_{klur\,vhqt} = \mathrm{Re}\left\{ \int_{-\frac{\tau_{sx}}{2}}^{\frac{\tau_{sx}}{2}} \int_{-\frac{\tau_{sz}}{2}}^{\frac{\tau_{sz}}{2}} S_{klur\,vhqt}\,dx\,dz \right\} \qquad (3.17)$$

$$S_{klur\,vhqt} = \frac{1}{2}\Big[E_{mz3klur\,vhq} H^{*}_{mx3klur\,vhq} - E_{mx3klur\,vhq} H^{*}_{mz3klur\,vhq} \Big]_{y=g} \qquad (3.16)$$

2.   MULTI-HARMONIC EQUIVALENT CIRCUIT

The multi-harmonic field model presented in chapter 3 section 1 takes into account the motor edge effects. These contribute to current asymmetry if the winding is supplied from a symmetrical voltage source. Due to this asymmetry the motor analysis should be carried out using a multi-harmonic equivalent circuit, separate for each phase winding.

In general, let a motor with an asymmetrical winding be supplied from an asymmetrical m phase voltage source as shown in Fig. 3-3. The voltage equations that correspond with the circuit in Fig. 3-3 are as follows:

$$V_1^{(1)} = E_o^{(1)} + \left(R_1^{(1)} + jX_{s1}^{(1)} \right) I_1^{(1)}$$

$$V_1^{(2)} = E_o^{(2)} + \left(R_1^{(2)} + jX_{s1}^{(2)} \right) I_1^{(2)}$$

$$\cdots\cdots\cdots\cdots\cdots\cdots\cdots\cdots\cdots\cdots \qquad (3.17)$$

$$V_1^{(m)} = E_o^{(m)} + \left(R_1^{(m)} + jX_{s1}^{(m)} \right) I_1^{(m)}$$

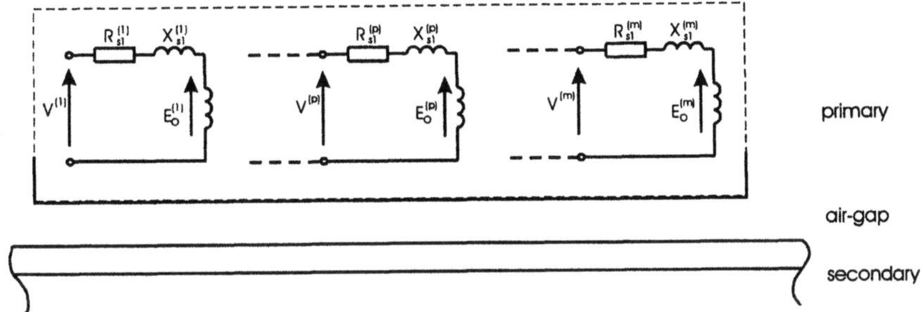

Figure 3-3. Circuit model of multi-phase winding of IM-2DMF

The circuit parameters R_1 and X_{s1} are resistances and leakage reactances of the primary phase windings. These can be determined using conventional methods.

$E_o^{(1)}, E_o^{(2)}, ..., E_o^{(m)}$ are the voltages induced in the phase windings by the magnetic fluxes linked with them. These magnetic fluxes are generated by m phase windings and are the sums of all harmonics expressed in the field model. The voltages are calculated as the sums of voltages $E_{o(n)}^{(m)}$ induced in n coils of each p phase:

$$\left. \begin{aligned} E_o^{(1)} &= E_{o(1)}^{(1)} + E_{o(2)}^{(1)} + \ldots + E_{o(n)}^{(1)} \\ E_o^{(2)} &= E_{o(1)}^{(2)} + E_{o(2)}^{(2)} + \ldots + E_{o(n)}^{(2)} \\ &\cdots\cdots\cdots\cdots\cdots\cdots\cdots\cdots\cdots\cdots\cdots \\ E_o^{(m)} &= E_{o(1)}^{(m)} + E_{o(2)}^{(m)} + \ldots + E_{o(n)}^{(m)} \end{aligned} \right\} \tag{3.18}$$

The $E_{o(n)}^{(m)}$ can be determined from the electric field component E_x in the layer $i = 2$ in the following way:

$$E_{o(n)}^{(m)} = \frac{j\omega T_{(n)}^{(m)}}{\sqrt{2}} \sum_k \sum_l \sum_u \sum_r \sum_v \sum_h \sum_q \sum_t \int_{-\frac{W}{2}}^{\frac{W}{2}}$$

$$\cdot \left\{ \begin{aligned} &\left[\left(E_{mx2klurvhrqt}^{(1)} \right)_{z=z_{(n)}^{(m)}} + \left(E_{mx2klurvhrqt}^{(1)} \right)_{z=z_{(n)}^{(m)}+\tau_z} \right] \\ &+ \left[\left(E_{mx2klurvhrqt}^{(2)} \right)_{z=z_{(n)}^{(m)}} + \left(E_{mx2klurvhrqt}^{(2)} \right)_{z=z_{(n)}^{(m)}+\tau_z} \right] \\ &+ \ldots\ldots\ldots\ldots\ldots\ldots\ldots\ldots\ldots\ldots\ldots\ldots\ldots\ldots\ldots\ldots \\ &+ \left[\left(E_{mx2klurvhrqt}^{(m)} \right)_{z=z_{(n)}^{(m)}} + \left(E_{mx2klurvhrqt}^{(m)} \right)_{z=z_{(n)}^{(m)}+\tau_z} \right] \end{aligned} \right\}_{\substack{y=0 \\ v=v_1}} dx \tag{3.19}$$

where:

$T^{(m)}_{(n)}$ – are the number of turns in the nth coil of mth phase winding

$z^{(m)}_{(n)}$ – are the positions of the coils on z axis.

The voltages $E^{(p)}_0$ induced in the pth phase can also be expressed in terms of stator currents, which generate the electromagnetic field whose components are as follows (see Appendix II):

$$E^{(p)}_{mxiklur\,vhqt} = I^{(p)} K^{'(p)}_i(y) F(x,z)$$

$$E^{(p)}_{mziklur\,vhqt} = -\frac{\tau_{zlr\,vhqt}}{\tau_{kux}} I^{(p)} K^{'(p)}_i(y) F(x,z)$$

$$B^{(p)}_{mxiklur\,vhqt} = -j\frac{\beta_i}{\omega}\frac{\tau_{zlr\,vhqt}}{\tau_{kux}} I^{(p)} K^{"(p)}_i(y) F(x,z) \qquad (3.19)$$

$$B^{(p)}_{myiklur\,vhqt} = \frac{\alpha_o}{\omega}\frac{\tau_{zlr\,vhqt}}{\pi} I^{(p)} K^{'(p)}_i(y) F(x,z)$$

$$B^{(p)}_{mxiklur\,vhqt} = -j\frac{\beta_i}{\omega} I^{(p)} K^{"(p)}_i(y) F(x,z)$$

According to this the *klurvhqt*th harmonic of the rms voltage induced in the pth phase winding is calculated as follows:

$$E^{(p)}_{oklurvhqt} = I^{(1)} j2 \sum_{c=1}^{c=n} \frac{\omega}{\sqrt{2}} T^{(p)}_c \int_{-\frac{W}{2}}^{\frac{W}{2}} K^{'(1)}_3\Big|_{y=0} \left[F\left(x, z^{(p)}_c\right) + F\left(x, z^{(p)}_c + \tau_z\right) \right] dx$$

$$+ I^{(2)} j2 \sum_{c=1}^{c=n} \frac{\omega T^{(p)}_c}{\sqrt{2}} \int_{-\frac{W}{2}}^{\frac{W}{2}} K^{'(2)}_3\Big|_{y=0} \left[F\left(x, z^{(p)}_c\right) + F\left(x, z^{(p)}_c + \tau_z\right) \right] dx$$

....

$$+ I^{(m)} j2 \sum_{c=1}^{c=n} \frac{\omega T^{(p)}_c}{\sqrt{2}} \int_{-\frac{W}{2}}^{\frac{W}{2}} K^{'(m)}_3\Big|_{y=0} \left[F\left(x, z^{(p)}_c\right) + F\left(x, z^{(p)}_c + \tau_z\right) \right] dx$$

$$(3.20)$$

or written in another form:

$$E^{(p)}_{oklurvhqt} = Z^{(1p)}_{oklurvhqt}I^{(1)} + Z^{(2p)}_{oklurvhqt}I^{(2)} + \ldots\ldots\ldots\ldots + Z^{(mp)}_{oklurvhqt}I^{(m)} \quad (3.21)$$

According to the above equations the motor equivalent circuit for p^{th} phase can be drawn as shown in the Fig. 3-4.

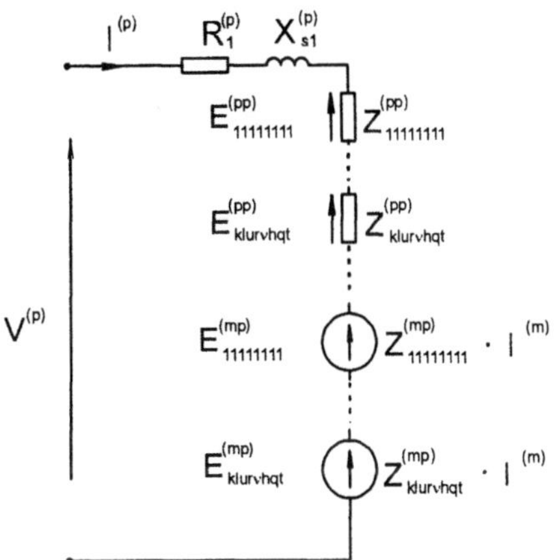

Figure 3-4. Multi-harmonic equivalent circuit of IM-2DMF for p^{th} phase

The EMFs $E^{(p)}$ are now shown as the voltage drops across the "self impedances" $Z^{(pp)}_{oklur\,vhqt}$ and "mutual impedances" $Z^{(mp)}_{oklurvhqt}$, where the voltages of the $klurvhqt^{th}$ harmonic depend on the currents flowing in the other windings.

The $klurvhqt^{th}$ harmonic of impedances might be rolled up into the resultant impedances expressed in the following way:

$$Z^{(pp)} = \sum_k \sum_l \sum_u \sum_r \sum_v \sum_h \sum_q \sum_t Z^{(pp)}_{klur\,vhqt} \qquad (3.22)$$

$$Z^{(mp)} = \sum_k \sum_l \sum_u \sum_r \sum_v \sum_h \sum_q \sum_t Z^{(mp)}_{klur\,vhqt} \qquad (3.23)$$

According to this, the equivalent circuit, which was used to farther calculations, is shown in Fig. 3-5.

It must be pointed out that the values of impedances depend on the rotor speed, so they have to be determined for the different speeds over the whole range the motor is to be operated.

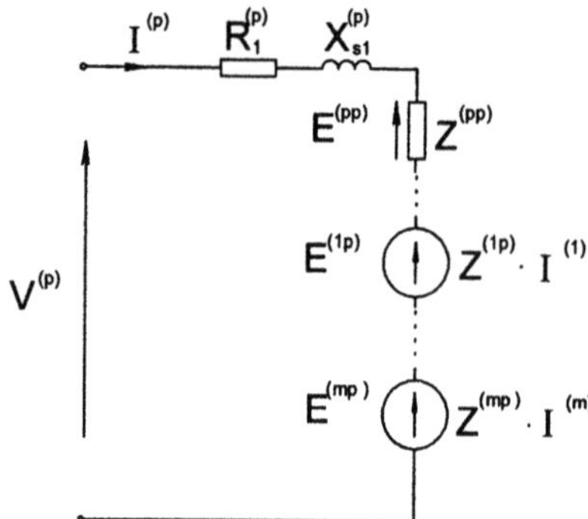

Figure 3-5. Multi-harmonic equivalent circuit of IM-2DMF for p^{th} phase with equivalent self- $Z^{(pp)}$ and mutual $Z^{(mp)}$ impedances

In the following chapters, calculation examples of modeling of induction motors with one- and two-degrees of mechanical freedom are presented.

Chapter 4

SIMULATION OF INDUCTION MOTORS WITH TWO DEGREES OF MECHANICAL FREEDOM

To illustrate the application of the theory derived in the preceding chapters the calculations of electromagnetic field and motor performance in steady-state conditions of a few types of motors with two degrees of mechanical freedom are presented. These are as follows:
- Rotary-linear induction motors; among them are:
  - twin-armature motor,
  - double-winding motor,
  - motor with rotating-traveling magnetic field,
- motor with spherical rotor,
- X-Y flat linear motor.

1.   ROTARY-LINEAR INDUCTION MOTORS

The common feature of this type of motor is the helical motion performed by the rotor. The speed of this motion consists of two components: rotary and linear speed. The share of each component differs during the motor operation. In two particular cases the rotor performs only one of them, moving either with rotary or linear motion. In the following sections a modeling and performance analysis of the three types of rotary linear motors mentioned above will be considered. Each motor type has a different construction, which determines its particular performance.

1.1   Twin-armature rotary-linear induction motor

The scheme of the motor construction was shown earlier in Fig. 1-2.The particular motor was built (Fig. 4-1) and the main dimensions are shown in Fig. 4-2.

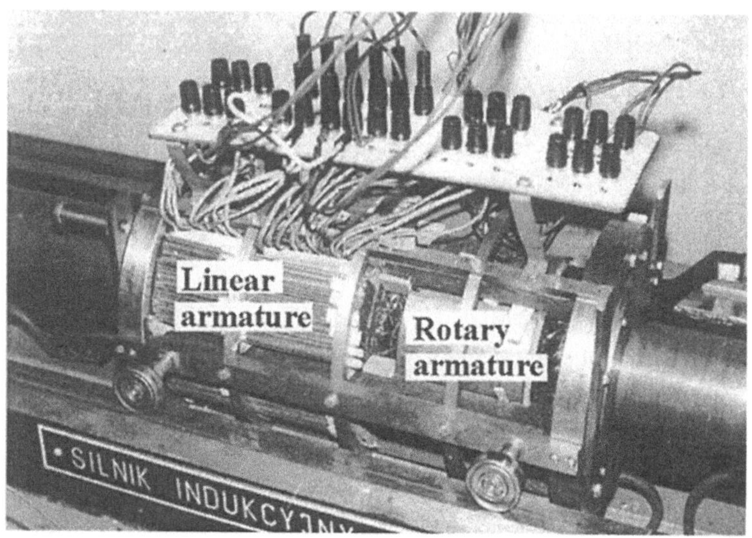

Figure 4-1. Laboratory model of twin-armature rotary-linear induction motor

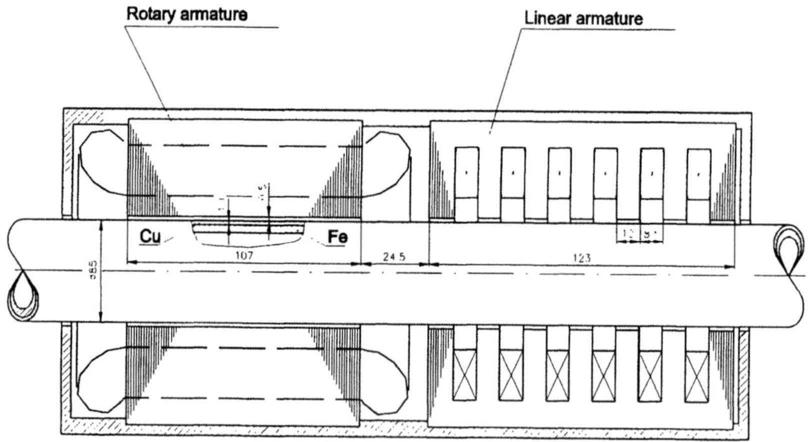

Figure 4-2.   Main dimensions of the laboratory model of twin-armature rotary-linear induction motor

As indicated, the two armatures produce the rotating and traveling magnetic fields that act on a common rotor. The result of that is the torque and thrust, which drive the rotor with a helical motion. If the two armatures are not too close to one another, the electromagnetic interaction between them is negligible. Thus the motor can be treated as a pair of two independent motors, rotary and linear, whose rotors are coupled stiffly by a mechanical clutch. However, since each armature has a finite length, some end effects will be expected. The influence of these effects on the magnetic field of both armature and further – on torque and thrust, is linked with the linear motion. Thus, there is an influence of linear motion on the "rotary motor", but the rotary motion does not influence the "linear motor", since both armatures are cylindrical. In consequence, though, the theoretical approach does not have to consider an electromagnetic link between two armatures, but it should take into account the influence of linear velocity on the performance of the "rotary motor".

-   **Computational model**

The computational field model of the motor that aims to determine the electromagnetic field in the motor is shown in Fig. 4-3. It is defined by the following assumptions:

a) Electromagnetic interaction between the two armatures is negligible, therefore the motor comprises a set of two independent motors: rotary and tubular linear, and their rotors are coupled stiffly and moving with a rotary linear motion.

b) Each motor comprises a 5-layer structure, whose layers are homogenous, isotropic and linear. They are defined by the following physical constants: stator iron core (layer 1) with magnetic permeability $\mu_1 = \infty$, electric conductivity $\gamma_1 = 0$; tooth zone (2) with μ_2, $\gamma_2 = 0$; air gap (3) with $\mu_3 = \mu_0$, $\gamma_3 = 0$; rotor conducting sheet (4) with μ_4, γ_4; rotor iron core (5) with μ_5, γ_5.

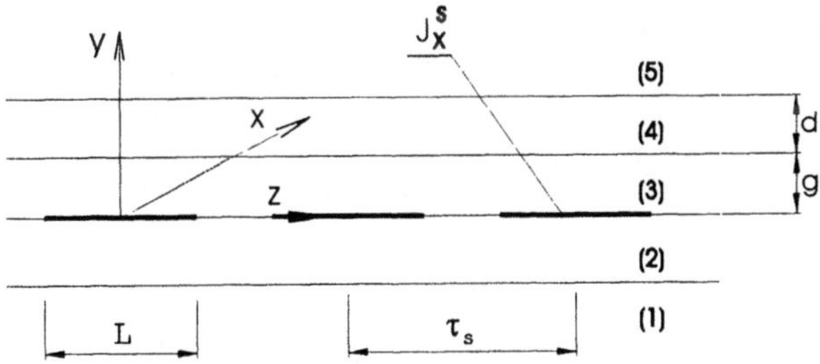

Figure 4-3. Computational model of rotary-linear induction motor (for a single armature of L length of twin armature motor)

c) The stator winding of each motor is represented by an infinitely thin current sheet attached to the layer 1, and its linear current density is defined as follows:

    − for the linear armature:

$$J_x(t,z) = \sum_v J_{xmv} \cos\left(\omega t - \frac{\pi}{\tau_{zv}} z\right) \tag{4.1}$$

    − for the rotary armature:

$$J_z(t,x) = \sum_v J_{zmv} \cos\left(\omega t - \frac{\pi}{\tau_{xv}} x\right) \tag{4.2}$$

where

$$J_{zmv}, J_{xmv} \quad - \text{ are amplitudes of the } v^{th} \text{ (phase) harmonic}$$

$$\tau_{zv}, \tau_{xv} \quad - \text{ are pole pitches of the } v^{th} \text{ harmonic}$$

$$\tau_{xv} = \frac{\tau_{x1}}{v}, \quad \tau_{zv} = \frac{\tau_{z1}}{v} \tag{4.3}$$

$$v = -1, 5, -7, 11, -13, \dots$$

d) The iron core of each armature is infinitely long in the z direction, but the current sheets have finite length L_r and L_l for the rotary and linear armatures respectively. Since a Fourier series method is used to solve the differential equations of the electromagnetic field, the current sheet of each armature is replaced by an infinite chain of identical current sheets (see Fig. 4-3 and also Fig. 3-2.), where the distance between adjacent sheets is $\tau_s - L$. The stator current densities of such current layers are described by the functions:

$$J_x(t,z) = \sum_l \sum_v J_{mxvl} \cos\left(\omega t - \frac{\pi}{\tau_{zv}} z\right) \cos\left(\frac{\pi}{\tau_{sxl}} z\right) \tag{4.4}$$

$$J_z(t,x,z) = \sum_l \sum_v J_{mzvl} \cos\left(\omega t - \frac{\pi}{\tau_{xv}} x\right) \cos\left(\frac{\pi}{\tau_{szl}} z\right) \tag{4.5}$$

After trigonometric transformation, Eqs. 4.4 and 4.5 take the form (see chapter 3):

$$J_x(t,z) = \sum_r \sum_l \sum_v \frac{1}{2} J_{mxvl} \exp\left[j\left(\omega t - \frac{\pi}{\tau_{zvlr}} z \right) \right] \qquad (4.6)$$

$$J_z(t,x,z) = \sum_r \sum_l \sum_v J_{mzvl} \exp\left[j\left(\omega t - \frac{\pi}{\tau_{xv}} x - \frac{\pi}{\tau_{zlr}} z \right) \right] \qquad (4.7)$$

where:

$$J_{mxvl} = \frac{4}{\pi} J_{mxv} \frac{1}{l} \sin\left(l \frac{L_l \pi}{2\tau_s} \right) \qquad (4.8)$$

$$J_{mzvl} = \frac{4}{\pi} J_{mzl} \frac{1}{l} \sin\left(l \frac{L_r \pi}{2\tau_s} \right) \qquad (4.9)$$

$$\tau_{zvlr} = \frac{\tau_{z1} \tau_s}{v\tau_s + rl\tau_{z1}}, \quad \tau_{zlr} = \frac{\tau_s}{rl} \qquad (4.10)$$

$l = -1, -3, -5, -7, \ldots$
$r = -1, 1$
L_l, L_r – are the lengths of the linear and rotary armatures.

Each vl^{th} harmonic of both current densities is represented by the two waves. For the linear armature these are two waves traveling along z axis with two different speeds. In the case of the rotary armature these are two rotating-traveling waves moving in the same circumferential direction, but in two opposite linear directions.

-　Motor performance

The main goal of modeling the motor is to analyze the motor performance. The calculation resulting from modeling should also illustrate the complex phenomena that do not exist in conventional motors, i.e. the edge effects, which are due to the helical motion of the rotor. The influence of this type motion can be illustrated in the magnetic field distribution in the air-gap as well as in the electromechanical characteristics determined at steady state conditions.

Such calculations were done for the particular motor shown in Fig. 4-1. They are presented in [34, 40, 46]. The data of this motor are as follows (see also Fig. 4-2):

Rotary armature

Armature length	$L_r = 0.107$ m
Pole pitch	$\tau_{x1} = 0.0675$ m
3-phase winding with the:	
– number of slots per pole per phase	$q_r = 2$
– linear current density amplitude	$J_{mz1} = 2738 \cdot I_z$ A/m

Linear armature:

Armature length	$L_1 = 0.123$ m
Pole pitch	$\tau_z = 0.0615$ m
3-phase winding with the:	
– number of slots per pole per phase	$q_1 = 1$
– linear current density amplitude	$J_{mx1} = 21996 \cdot I_x$ A/m

Rotor

Rotor length	$L_{rot} = 2.0$ m
Rotor diameter	$D_r = 0.086$
Copper sheet thickness	$d = 0.0011$ m
Air gap	$g = 0.0005$ m
Conductivity of the copper sheet	$\gamma_3 = 57 \cdot 10^6$ S/m
Conductivity of the rotor iron layer	$\gamma_4 = 5.9 \cdot 10^6$ S/m

The calculations of magnetic field distributions were carried out for the supply frequency f = 50 Hz, and armature currents: $I_r = 3.24$ A, and $I_1 = 3$ A. The results are shown in Fig. 4-4 in the form of magnetic flux density distributions in the air gap on the rotor surface along both motor armatures.

To illustrate the influence of finite armature length, the characteristics were determined at two rotor linear speeds and compared with the characteristics of infinity long armatures ($L = \infty$) where no end effects occur. The significant deformation of characteristics illustrates the influence of finite armature length at high linear speed. The magnetic field is damped at entry edge of both armatures, while it is sustained at the exit edges.

The mechanical characteristics were determined first at constant current supply, then at constant voltage using the motor equivalent circuit presented in chapter 2. The results are published in detail in [10, 11, 45]. Here in Figs. 4-5, 4-6 and 4-7 characteristics calculated at constant supply voltage are presented. Fig. 4-5 shows the family of characteristics of torque and linear force vs. rotary slip determined for rotary armature at different linear speeds. The higher this speed is the more deteriorated the torque and the higher the linear force that breaks the rotor.

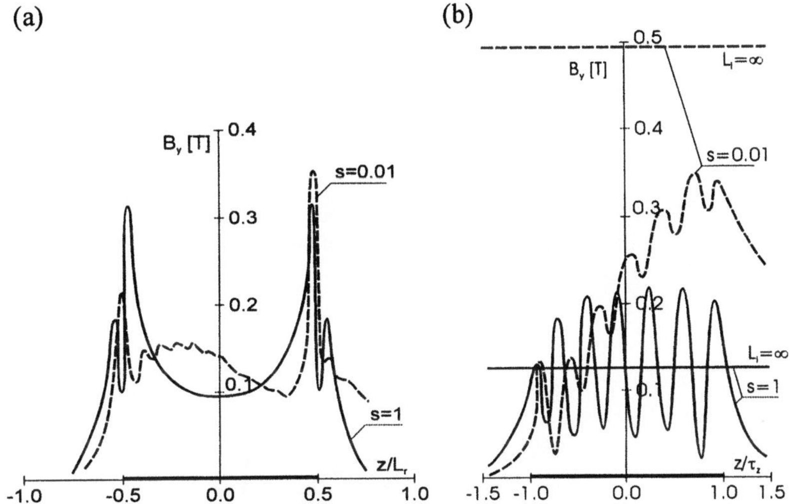

Figure 4-4. Magnetic flux density distributions on the rotor surface of twin-armature rotary-linear motor: (a) along rotary armature, (b) along linear armature at n = 0

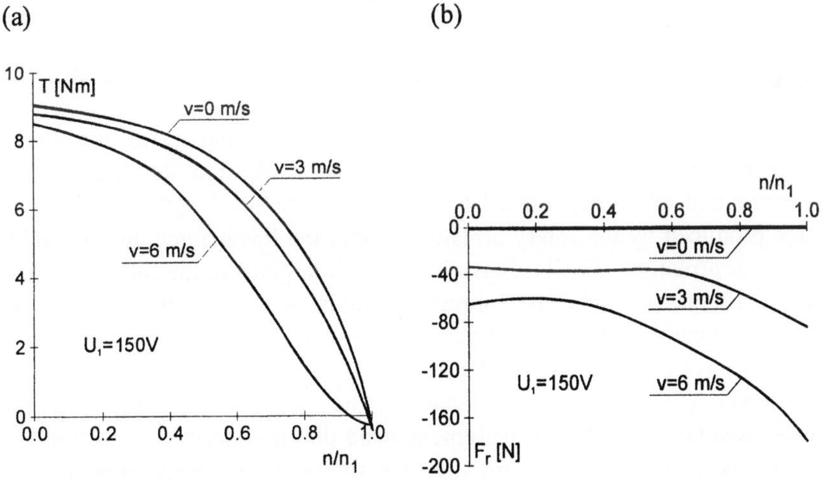

Figure 4-5. Characteristics of: (a) torque *T* and (b) linear force *F_r* vs. rotary speed *n* produced by the rotary armature at different linear speeds *v_z* of the rotor

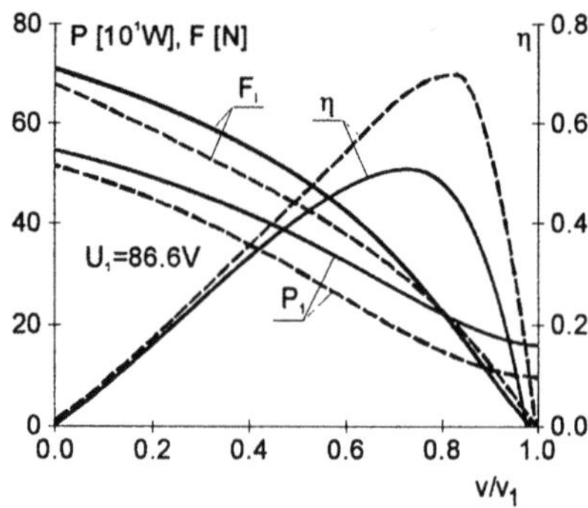

Figure 4-6. Force F_1, input power P_1 and efficiency η vs. linear speed v characteristics calculated for infinitely long (----) and of finite length (——) linear armature

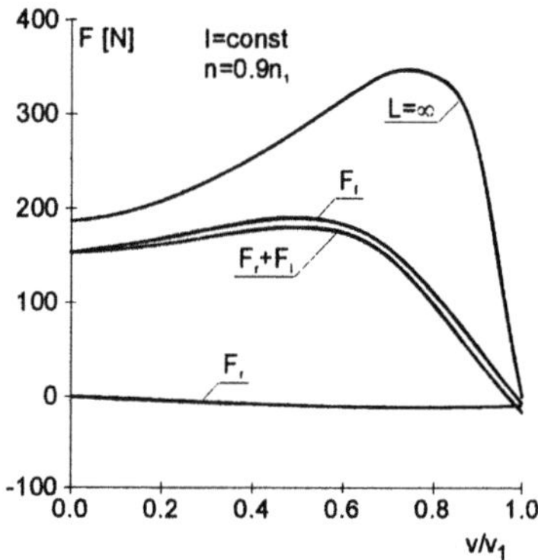

Figure 4-7. Characteristics of linear force produced by the motor vs. linear slip

Fig. 4-6 shows mechanical characteristics calculated for the linear armature with and without end effects. They also illustrate the deterioration of thrust and efficiency caused by the finite armature length. The braking force produced by the rotary armature lowers the linear force developed by the linear armature. This is shown in Fig. 4-7. In general, the finite armature length affects not only the torque and thrust but also contributes to an increase of currents and electric power losses and a decrease of power factor. This is shown in more detail in [16, 20, 55].

To verify the accuracy of modeling the motor, the laboratory motor model was tested on the measurement stand that is described in section 1.4 and in the references [32, 36]. Fig. 4-8 shows the magnetic flux density distributions along both armatures measured and calculated on the rotor surface. In Fig.4.9 some of the electromechanical characteristics are presented, measured and calculated at constant supply voltage.

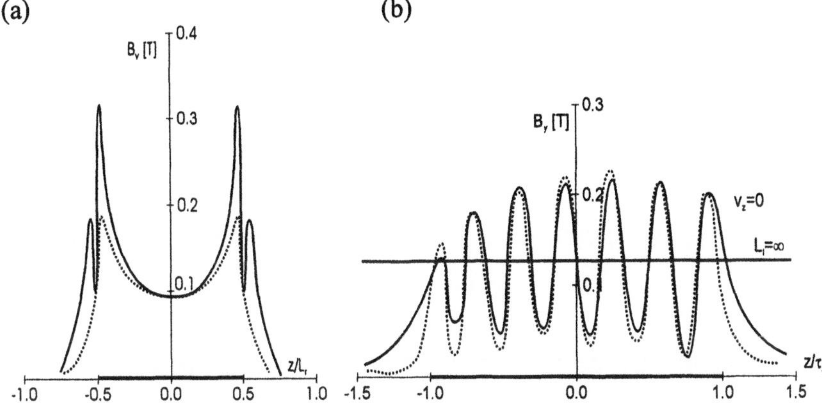

Figure 4-8. Experimental (·····) and simulated characteristics of magnetic flux density distribution on the surface of the rotor of twin-armature rotary-linear motor: (a) for rotary armature, (b) for linear armature at zero rotor speed

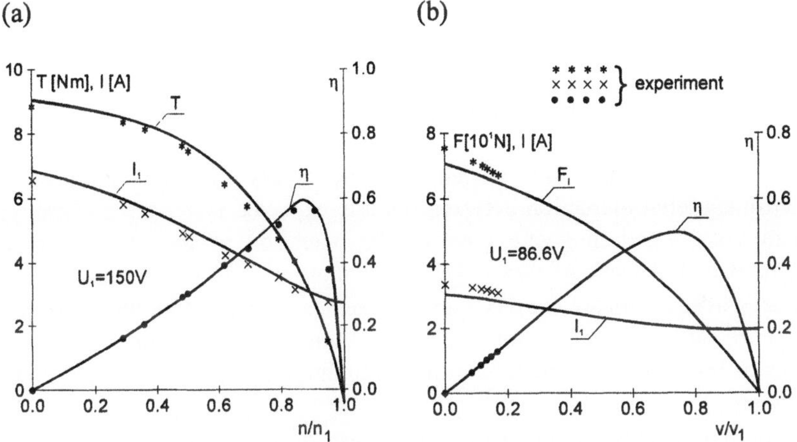

Figure 4-9. Experimental and simulated characteristics determined: (a) for the rotary armature (torque T, current I and efficiency η vs. rotary speed n), (b) for the linear armature (linear force F_l, current I and efficiency η vs. linear speed v)

Due to the short rotor high linear speed has not been reached during the measurements. The discrepancies between the characteristics obtained from the measurements and simulation are not significant. This indicates that the mathematical model describes the motor with relatively good accuracy.

1.2   Double-winding rotary linear motor

The double-winding rotary-linear motor is a modification of the twin-armature motor. It has similar rotary and linear windings put together in the same armature core in such a way that one winding overlaps another (Fig.4.10).

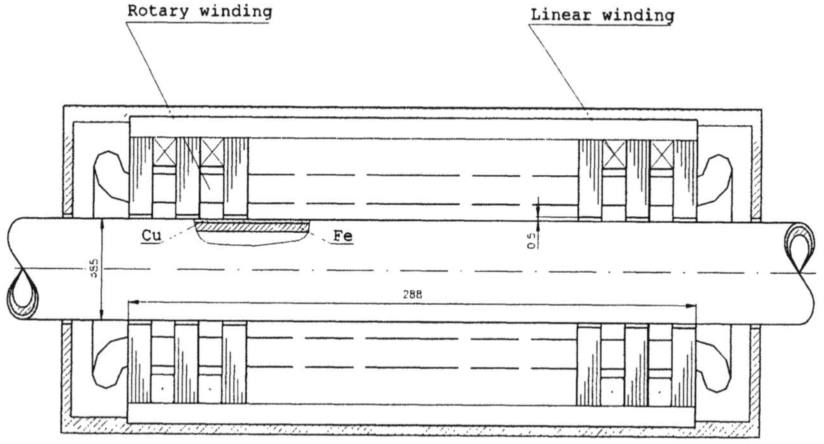

Figure 4-10. Scheme of double-winding rotary-linear induction motor

Since the windings are placed perpendicularly to one another no electromagnetic interaction between them is expected under the assumption of an unsaturated magnetic circuit. The operation of the motor can be regarded as the work of two independent motors, both rotary and tubular-linear, whose rotors are coupled rigidly together. In the real motor, where the iron core can be partially saturated by the cumulative interaction of magnetic fields of both windings the mutual magnetic interaction should be taken into account.

The mathematical model of the motor assumes the linearity of all motor elements. In this case the electromagnetic properties will not differ from that of the twin-armature motor. The only difference that can occur lies in the magnetic flux density distribution on the rotor surface. The calculations of the electromagnetic field should be performed jointly for two linear current densities.

The laboratory motor model shown in Fig. 4-11 is an object of analysis. Its constructional data are as follows:

Rotary winding:

Pole-pitch	$\tau_x = 0.0675$ m
Tooth pitch	$\tau_{tx} = 0.01125$ m
Slot opening	$b_{sx} = 0.0028$ m

Number of slots per pole per phase $q_r = 2$
Number of wires in the slot $w_r = 60$
Linear current density of the first phase harmonic (ν)
(no tooth harmonics are considered):

$$J_{mz1} = \frac{6\sqrt{2}\,I_r w_r}{\pi \tau_{tx}} \sin\left(\frac{\pi}{6}\right)$$

Linear winding:
Pole-pitch $\tau_z = 0.072$ m
Tooth pitch $\tau_{tz} = 0.024$ m
Slot opening $b_{sz} = 0.012$ m
Number of slots per pole per phase $q_l = 1$
Number of wires in the slot $w_l = 210$
Linear current density of the first phase harmonic (ν)
(no tooth harmonics are considered):

$$J_{mx1} = \frac{6\sqrt{2}\,I_l w_l}{\pi \tau_{tz}} \sin\left(\frac{\pi}{6}\right)$$

Armature core length $L = 0.288$ m
Rotor – the same as for twin-armature motor

Figure 4-11. Laboratory model of double-winding rotary-linear induction motor

The calculations of the electromagnetic field and motor performance were done for the supply frequency f = 50Hz. The magnetic flux density distributions determined on the rotor surface along the stator length (z axis) at different rotor speeds, when two windings were supplied, are shown in Fig. 4-12.

In the middle of the stator length the maximums can be spotted. It is the result of interference of two opposite magnetic poles of rotating and traveling fields. The position of these deformations, though stable on the rotor surface, differs on the rotor circumference with respect to the longitudinal axis if two windings are supplied with the same frequency. These deformations will be moving if the windings are supplied with different frequencies.

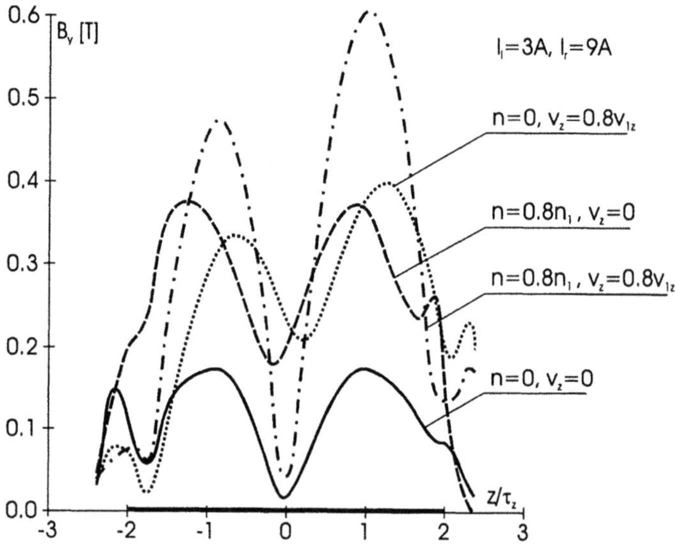

Figure 4-12.   Magnetic flux density distribution on the rotor surface of double-winding rotary-linear motor at different rotary n and linear v_z speeds

To make the characteristics more readable, no tooth harmonics of linear and rotary windings were included. These harmonics significantly deteriorate the motor performance causing a decrease of torque and linear force and an increase of power losses in the rotor. This is illustrated in Fig. 4-13 where rotary and linear forces vs. rotary speed characteristics, determined for the rotary winding at different linear speeds are presented.

Although, the performance of twin-armature and double-winding motors are similar, the application of the first motor mentioned is more advisable, not only because of better performance, but also due to the easier manufacturing process.

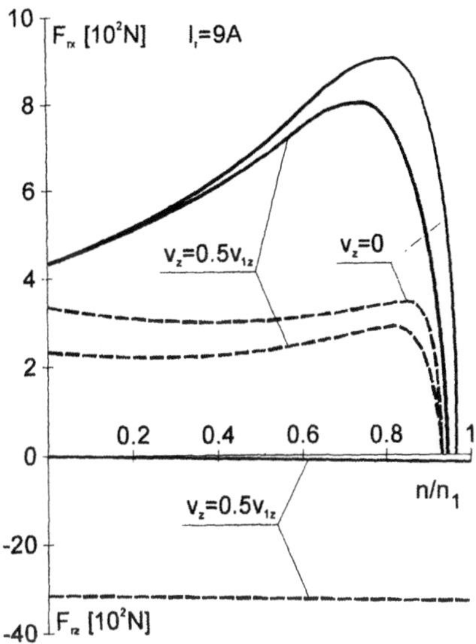

Figure 4-13.   Mechanical characteristics of rotary force F_{rx} and linear force F_{rz} vs. rotary speed n determined for rotary winding of double-winding rotary-linear motor at different linear speeds v_z

To verify the mathematical model of the motor, a test was carried out on the motor laboratory model. Fig. 4-14 shows the magnetic flux density distribution measured in the air-gap on the rotor surface, along the motor axis. The measurements that show the influence of the linear winding slotting are compared with simulated characteristic calculated without inclusion of stator slots. More details on this motor are presented in [41].

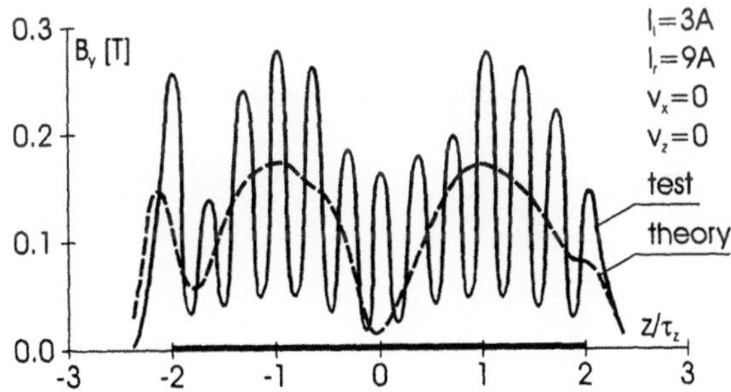

Figure 4-14.   Experimental and theoretical characteristics of magnetic flux density distribution on the rotor surface of double-winding rotary-linear motor when two windings are supplied

1.3   **Rotary-linear motor with rotating-traveling field**

-   **Construction and operation**

The two versions of rotary-linear motors discussed in the previous sections allow the development of the helical rotor motion due to the generation of two magnetic fields by two independent windings: rotating and traveling. The motor discussed in this section develops the resultant magnetic field, which moves helically. The scheme of the motor construction is shown in Fig. 4-15.

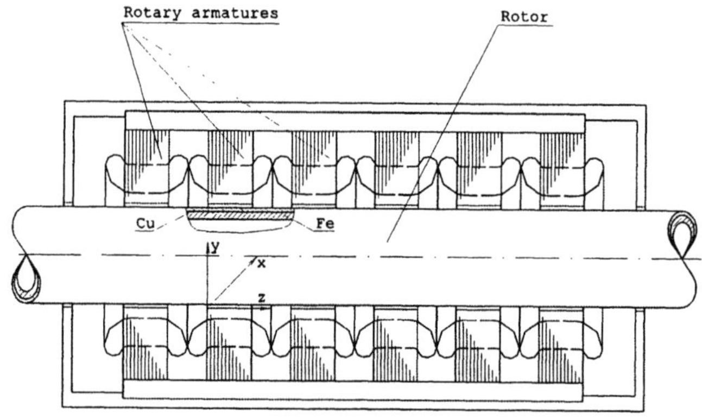

Figure 4-15.   Construction scheme of rotary-linear induction motor with rotating-traveling field

The stator consists of several armatures, built similarly to the one of a conventional rotary induction motor. Each armature produces a rotating magnetic field. If all of them rotate with the same phase (Fig. 4-16.a) the resultant field is a rotating one. When there is a shift between the adjacent rotating fields by the same angle (Fig. 4-16.b), the resultant field moves helically.

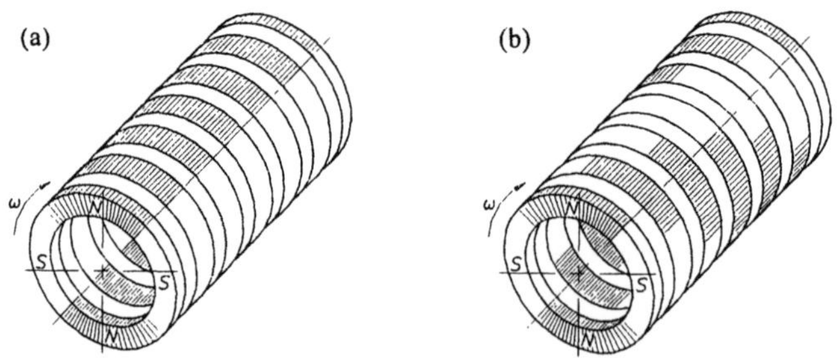

Figure 4-16. Generation of (a) rotating and (b) rotating-traveling fields

In order to change the motion direction of this field two methods can be applied: mechanically – by turning round the armatures, and electrically – by changing the connections of winding terminals of all armatures. In the latter case the direction of the field motion changes stepwise. The direction of force affecting the rotor changes accordingly the field motion direction.

A modification of this type of motor is the construction with linear armatures placed symmetrically in parallel around the rotor surface. The performance of this type of motor is discussed in [32].

-　Computational model

The general mathematical field model of the motor with rotating-traveling field has been described in chapter 2 and chapter 3. The linear current density of the stator that produces the rotating magnetic field has been written in general form without referring to the particular construction. Due to that, it is necessary to describe the current density for this particular motor.

Now, considered is the motor that was built as a prototype, which has six identical rotary armatures. The stator armatures are supplied from a 3-phase source and they are connected in such a way that their current linear densities are displaced from one another by the angle *π/3 rad* as shown in Fig. 4-17.

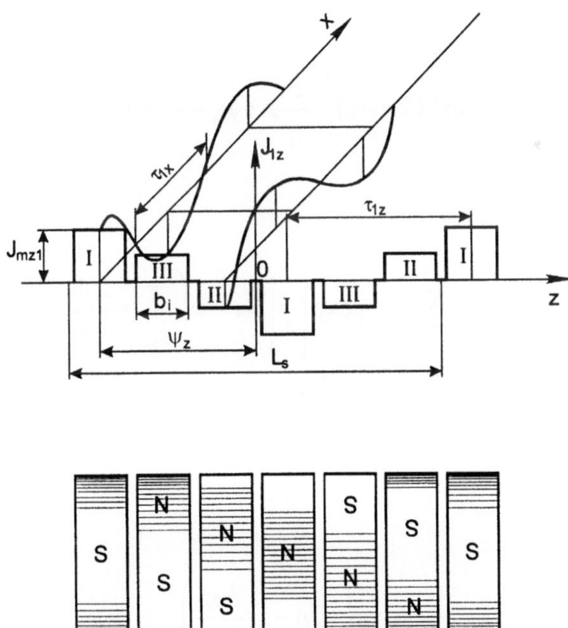

Figure 4-17. Distribution of linear current densities of rotary armatures

The current density of each armature is described by the following function:

$$J_z(t,x) = \sum_{v} J_{zmv} \cos\left(\omega t - \frac{\pi}{\tau_{xv}} x\right) \tag{4.11}$$

If every third armature is considered as a pair then the current densities of the 3 pairs are expressed in the following way:

– 1-st pair (Fig. 4-17)

$$J_z^{(I)}(t,x,z) = \sum_{n}\sum_{v} J_{mzvn} \exp\left[j\left(\omega t - v\frac{\pi}{\tau_{1x}} x\right)\right] \\ \cdot \cos\left[(2n-1)\left(\frac{\pi}{\tau_{1z}} z + \psi_z\right)\right] \tag{4.12}$$

– 2-d pair

$$J_z^{(II)}(t,x,z) = \sum_{n}\sum_{v} J_{mzvn} \exp\left[j\left(\omega t - v\left(\frac{\pi}{\tau_{1x}} x + \frac{2}{3}\pi\right)\right)\right] \\ \cos\left[(2n-1)\left(\frac{\pi}{\tau_{1z}} z + \psi_z - \frac{2}{3}\pi\right)\right] \tag{4.13}$$

– 3-d pair

$$J_z^{(III)}(t,x,z) = \sum_{n}\sum_{v} J_{mzvn} \exp\left[j\left(\omega t - v\left(\frac{\pi}{\tau_{1x}} x + \frac{4}{3}\pi\right)\right)\right] \\ \cos\left[(2n-1)\left(\frac{\pi}{\tau_{1z}} z + \psi_z - \frac{4}{3}\pi\right)\right] \tag{4.14}$$

where:

$$J_{mzvn} = \frac{4}{\pi} J_{mzv} \frac{1}{(2n-1)} \sin\left((2n-1)\frac{b_i\pi}{2\tau_{z1}}\right) \tag{4.15}$$

n = 0, 1, 2, 3, ...
b_i – length of the rotary armature.
The resultant current density is

$$J_z = J_z^{(I)} + J_z^{(II)} + J_z^{(III)} \tag{4.16}$$

$$J_z = \sum_m \sum_v J_{mzvm} \exp\left[j\left(\omega t - v\frac{\pi}{\tau_{1x}}x - m\left(\frac{\pi}{\tau_{1z}}z + \psi_z\right)\right)\right] \tag{4.17}$$

where:

$$J_{mzvm} = \frac{6}{\pi m}\sin\left(m\frac{b_i\pi}{2\tau_{1z}}\right)J_{mzv} \tag{4.18}$$

m = 1, -5, 7, -11, 13, …..
ψ_z = displacement angle of the 1st harmonic related to the coordinate
     system (Fig. 4-17)
If the finite length of the stator L_s is taken into account, then:

$$J_z = \sum_l \sum_r \sum_m \sum_v \frac{1}{2}J_{mzvml} \exp\left[j\left(\omega t - \frac{\pi}{\tau_{xv}}x - \frac{\pi}{\tau_{zmlr}}z + m\psi_z\right)\right] \tag{4.19}$$

where:

$$J_{mzvml} = \frac{4}{\pi l}\sin\left(l\frac{L_s\pi}{2\tau_{sz}}\right)J_{mzvm} \tag{4.20}$$

$$\tau_{zmlr} = \frac{L_s\tau_{sz}}{\tau_{sz} + rlL_s} \tag{4.21}$$

l = 1, 3, 5, 7, …
r = 1, -1.
The linear current density expressed by function 4.19 takes into account
harmonics that come from the phase groups (v), finite length of each rotary

armature (m) and finite length of the stator (l). Index (r) is related to the forward and backward waves of vml^{th} harmonic.

Relation 4.19 is the basis for further calculations of the electromagnetic field using the formulae derived in chapter 2.

The motor that was analyzed is shown in Fig. 4-18 and its data are as follows:

Stator

Number of rotary armatures	$N = 6$
Internal diameter of the armatures	$D_s = 0.086$ m
Length of the rotary armature	$b_i = 0.0305$ m
Stator length	$L_s = 0.438$ m
The distance between rotary armatures	$b_a = 0.0425$ m
The number of poles	$p = 4$
Pole pitch	$\tau_x = 0.0675$ m
Tooth pitch	$\tau_t = 0.0113$ m
Number of slots per pole per phase	$q = 2$
Number of wires in slot	$w = 30$

Linear current density of the rotary armature

$$J_{mzv} = \frac{6\sqrt{2}Iw}{\pi\tau_t v}\sin\left(v\frac{\pi}{6}\right)\cos\left(v\frac{\pi}{12}\right)$$

Rotor – the same as for twin-armature motor

Figure 4-18.   Laboratory model of the rotary-linear induction motor with rotating-magnetic field

The construction of the motor prototype allows the change of the mutual positions of the adjacent rotating magnetic field mechanically (turning the rotary armatures) and electrically (by changing the connections of armature windings). The test and simulation results of electromagnetic field distribution and performance characteristics were published in [40, 44].

In Fig. 4-19 the force-speed characteristics determined for an infinitely

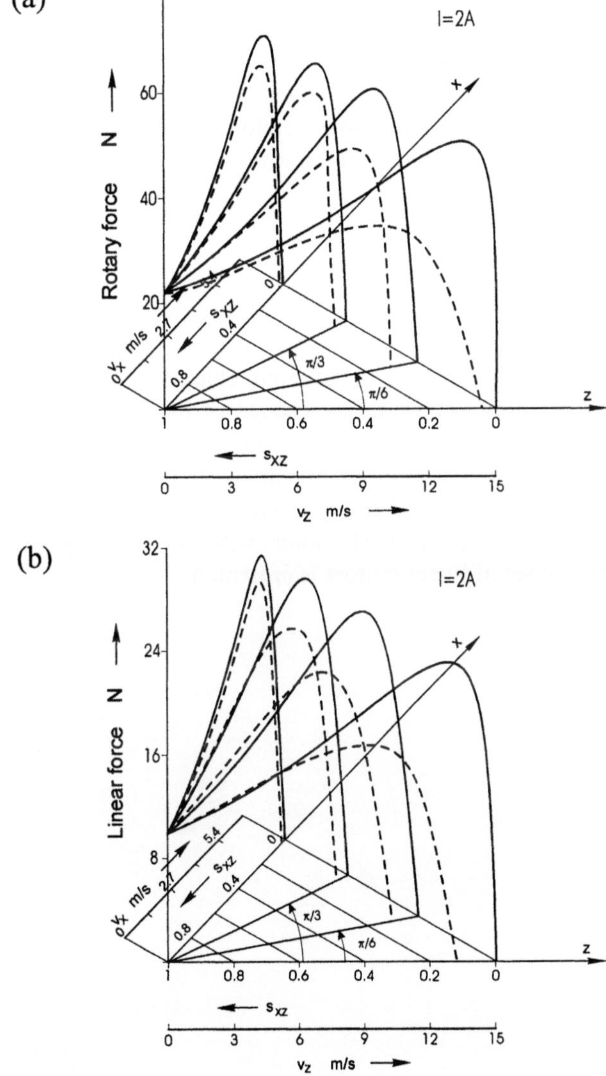

Figure 4-19. Characteristics of (a) rotary and (b) linear forces vs. rotary-linear slip s_{xz} calculated at constant supply current

long motor (no end effects are included) and the motor with finite length, operating in different motion directions are presented. They were calculated at 50 Hz frequency and constant supply current $I = 2$ A, when the shift angle between adjacent rotating magnetic field was equal to $60°$. The share of end effects increases significantly if the linear speed component increases. It is much smaller when the motor moves with rotary motion only ($v_z = 0$).

One of the unique features of the motor with rotating-traveling field is the linear relationship between linear (v_z) and rotary (v_x) speed components at constant rotor slip (see Eqn. 2.13). In the case of the induction version of the motor it is very difficult to verify the linear relationship experimentally due to the problem of keeping constant slip during the motor operation. It is easier to do it with the synchronous version of the motor. For this purpose the test was carried out on the motor with the reluctance rotor shown in Fig. 4-20.

According to the relation

$$v_z = \frac{v_{1z}}{n_1}\left(n_1 - n\right),$$
(4.22)

obtained from Eqn.2.13 at $s_{xz} = 0$, the minor changes around the synchronous rotary speed n_1 should result in the change of the linear motion direction. The experiment proved this conclusion. The change of rotary speed above and below its synchronous value caused an oscillatory motion of the motor.

A study on steady-state stability conditions of the rotary-linear motors can be found in [43]. In [33] motor with the rotating-traveling field generated by the set of linear motors is presented.

Figure 4-20. Synchronous version of rotary-linear induction motor with a reluctance rotor

1.4   Measurement stand for rotary-linear motor tests

Measurements of the motors with two degrees of mechanical freedom are more complex than those for rotary or linear motors because they require loading and simultaneously recording of the mechanical data in two motion directions. This also influences the complexity of the measurement stand for testing of rotary-linear motors. This-type of measurement stand was built and is shown schematically in Fig. 4-21.

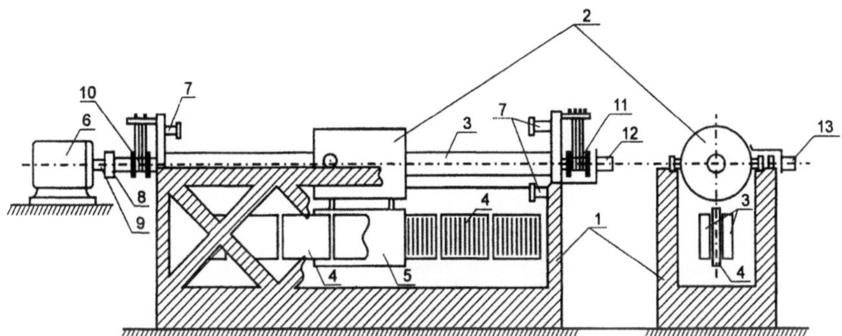

Figure 4-21.   Scheme of the measurement stand for rotary-linear motor tests: 1 – frame, 2 – RLIM, 3 – rotor, 4 – flat linear induction motors (LIM), 5 – aluminum plate (secondary part of LIM), 6 – DC generator, 7 – end springs, 8 – clutch, 9 – torsion shaft (for torque measurement), 10 and 11 – slip ring heads with brushes, 12 and 13 – tacho-generators to measure rotary and linear speeds respectively

The entire construction allows loading and measuring of mechanical quantities independently in both linear and rotary directions. The stator suspended on steel rolls (wheels) is moving linearly on two steel rails and is loaded by the flat linear motors suspended under the rotary-linear motor. The rotor that can move only in the rotary direction is loaded by the dc generator. Due to the limited length of the linear path for the moving armature (2 m long), there are end springs mounted at both ends. They allow, together with end switches, an oscillatory armature movement.

The test stand was equipped with a number of sensors and Fig. 4-22 shows their distribution schematically. The sensors allow measurement of the following quantities:
- rotary and linear speeds – by means of tachogenerators connected to stator and rotor,
- stator displacement – by means of potentiometer attached to the stator,
- torque and thrust – by means of strain gauges
- rotor temperature – thermo-resistive temperature sensor
- magnetic flux density in the air gap – by means of Hall sensors,

– circumferential and axial components of electric field intensities – by
   means of measurement coils.

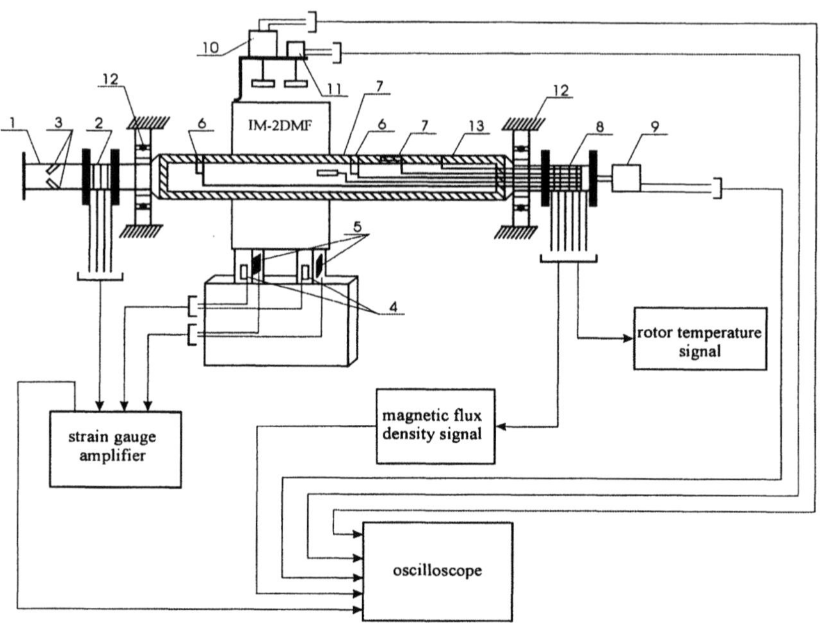

Figure 4-22.   Scheme of the measurement stand showing the distribution of sensors: 1 –
               torsion shaft for torque measurement, 2 and 8 – slip ring heads, 3 – strain
               gauges for torque measurement, 4 – strain gauges for measurement of side
               forces, 5 – strain gauges for measurement of thrust, 6 – temperature sensor, 7 –
               Hall sensors, 9 – tachogenerator to measure the rotary speed, 10 –
               tachogenerator for measurement of linear speed, 11 – potentiometer for
               measurement of stator position, 12 – bearings, 13 – coils for measurement of
               tangential component of magnetic flux densities

The signals measured on the rotor surface are transmitted by means of
wires placed inside the rotor to the slip ring head and next via brushes to the
measurement equipment. The measurement stand shown in Fig. 4-23 also
allows testing flat and tubular linear motors. The test stand is described in
more detail in [36].

Research on rotary-linear motors was also carried out by G. Kaminski at
Warsaw University of Technology [23, 25], J.J. Cathey and M. Rabiee from
Kentucky University [2, 3, 59, 60], and by other researches [21, 22, 27, 54].

Figure 4-23. Measurement stand for testing of rotary-linear motors

2. INDUCTION MOTOR WITH SPHERICAL ROTOR

The induction motor with spherical rotor shown schematically in Fig. 1-3 is a counterpart of the twin-armature rotary-linear motor. Another version of the motor with spherical rotor with two windings, placed perpendicularly to each other in one core would be the counterpart of the double-winding rotary-linear motor. Both versions of the motor with spherical rotor are expected to differ in their electromechanical properties similar to their cylindrical counterparts. As an example for modeling, the version with two armatures is to be considered.

In the calculation model the symmetrical position of both armatures on the surface of the spherical rotor is assumed. One of the armatures produces a magnetic field traveling meridionally (in x direction), the other generates a field moving in the equatorial direction (z direction). Modeling in Cartesian's coordinate system errors are expected, which result from the assumption that there is the same speed of all points on the rotor surface with respect to armature. In the real motor, points placed aside of the armature center will have slightly smaller speeds. This error will be larger for a greater armature surface (with respect to the entire rotor surface).

The calculation model does not differ from the one derived in chapters 2 and 3. To illustrate the application of this model to the analysis of the real object, the motor of the following constructional characteristics is considered:

The two armatures are identical and placed perpendicularly to one another. Their data are as follows:

Armature surface (length L and width W) $L \cdot W = (0.2 \cdot 0.2)\ \text{m}^2$
Pole pitch $\tau = 0.05\ \text{m}$
Tooth pitch $\tau_t = 0.0167\ \text{m}$
Number of slots per pole per phase $q = 1$
Number of wires in the slot $w = 210$
Linear current density:

$$J_{mv} = \frac{6\sqrt{2}Iw}{\pi\tau_t v}\sin\left(v\frac{\pi}{6}\right)$$

air-gap $g = 1\ \text{mm}$
rotor is an aluminum ball, empty inside and
the thickness of its wall $d = 10\ \text{mm}$
rotor conductivity $\gamma_{Al(70°C)} = 31.82\cdot10^6\ \text{S/m}$

Calculations were carried out for the frequency $f = 50$ Hz. The simulation results of magnetic flux density distribution along the x (direction of the field traveling) and z directions determined at various motion directions defined by angle α are shown in Fig. 4-24. The small deformations of the magnetic field in transverse (z) distribution illustrate the negligible influence of edge effects caused by the rotor motion. This influence would be greater in the case of back iron in the rotor structure, as it was in rotary-linear motors. The influence of finite dimensions of the armature on motor performance at different direction of the rotor motion is visible in the force speed characteristics presented in Fig. 4-25. When the rotor moves aslant with respect to the field direction then the driving force F_x decreases and the braking force F_z occurs.

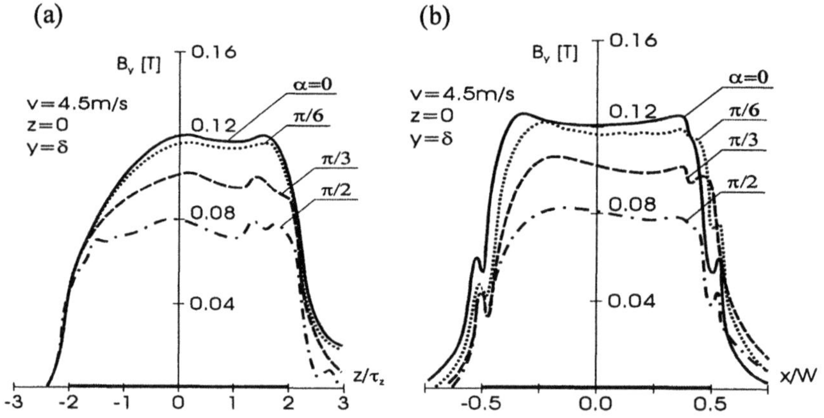

Figure 4-24.   Distribution of magnetic flux density on the surface of the rotor with spherical rotor: (a) along z axis (direction of traveling field motion), (b) along x axis

The rotor motion is the results of action of two armatures. Fig. 4-26 shows the characteristic of force produced by two armatures when they are supplied from the same voltage source. The direction of the resultant force is the same as the rotor motion only at the angle $\alpha = 45^o$. In the case of the same loading in all motion directions the rotor will move according to the resultant force. Its direction and the direction of rotor motion can be controlled by the change of supply voltages and frequencies.

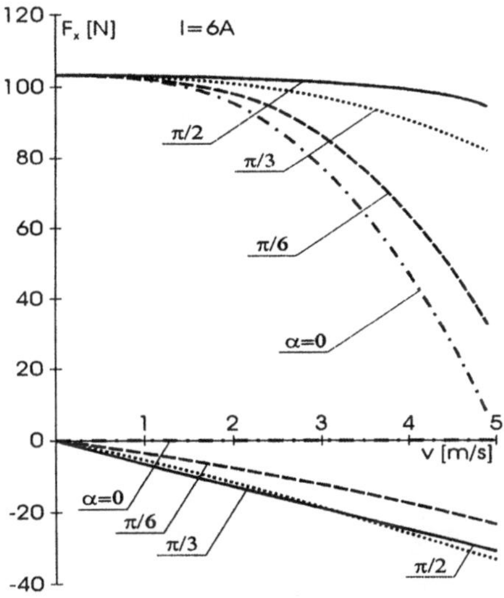

Figure 4-25. Characteristics of force components vs. tangential speed component v determined for a single armature of an induction motor with spherical rotor

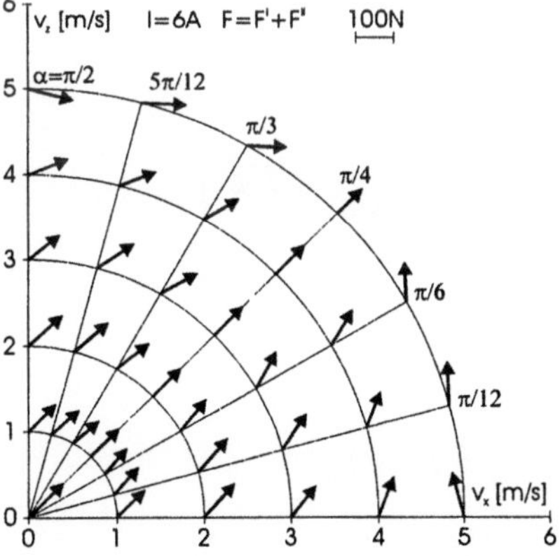

Figure 4-26. Diagram of resultant force at different rotor speed directions when the two armatures of the motor with spherical rotor are supplied from the same source

The operation of the motor with spherical rotor is described in more detail in [40]. A Research on spherical motors was carried out also by G. Kaminski and was published in [24, 26].

3. X-Y LINEAR INDUCTION MOTOR

The calculation model of an induction motor with two degrees of mechanical freedom defined in chapter 3 refers to the X-Y motor shown schematically in Fig. 1-1. The only difference is in the linear current density, which in this version of the motor is produced by two windings. If the finite width and length of the armature is considered, the windings and phase currents are symmetrical, and no slot harmonics are taken into account, the current densities of both windings are described by the following functions:

$$J_x(t,z) = \sum_l \sum_k \sum_u \sum_r \sum_v \frac{1}{4} J_{mxklv} \exp\left[j\left(\omega t - \frac{\pi}{\tau_{xku}} x - \frac{\pi}{\tau_{zlrv}} z \right) \right] \quad (4.23)$$

$$J_z(t,z) = \sum_l \sum_k \sum_u \sum_r \sum_v \frac{1}{4} J_{mzklv} \exp\left[j\left(\omega t - \frac{\pi}{\tau_{xlrv}} x - \frac{\pi}{\tau_{zku}} z \right) \right] \quad (4.24)$$

where:

$$J_{mxklv} = \frac{16}{\pi^2 kl} \sin\left(k\frac{W\pi}{2\tau_{sx}} \right) \sin\left(l\frac{L\pi}{2\tau_{sz}} \right) J_{mxv} \quad (4.25)$$

$$J_{mzklv} = \frac{16}{\pi^2 kl} \sin\left(k\frac{W\pi}{2\tau_{sx}} \right) \sin\left(l\frac{L\pi}{2\tau_{sz}} \right) J_{mzv} \quad (4.26)$$

$$\tau_{xku} = \frac{\tau_{sx}}{uk}, \quad \tau_{zvlr} = \frac{\tau_{zv}\tau_{sz}}{lr\tau_{zv} + \tau_{sz}} \quad (4.27)$$

$$\tau_{xvlr} = \frac{\tau_{xv}\tau_{sx}}{lr\tau_{xv} + \tau_{sx}}, \quad \tau_{zku} = \frac{\tau_{sz}}{uk} \quad (4.28)$$

For the current layer of the primary defined by the above equations further analysis should be carried out by using the equations derived in chapter 3. Due to the four armature edges the electromechanical properties of the motor and existing edge effects will be similar to the ones of the motor with spherical rotor presented in section 2.

Chapter 5

SIMULATION OF INDUCTION MOTORS WITH ONE DEGREE OF MECHANICAL FREEDOM

Among induction machines with one degree of mechanical freedom the cylindrical rotary motors are the most trivial case to simulate. The category of machines where more complex phenomena take place is motors for special purposes. That category contains the following motors that will be analyzed in this chapter:
- linear induction motors,
- disc-type induction motors,
- MHD machines.

1. LINEAR INDUCTION MOTORS

There are a number of different types of linear motors. One of them is a flat linear motor with short primary where electromagnetic phenomena seem to have the most complex form due to the finite length and width of the primary. These phenomena are even more complex if the secondary is moving in a different direction with respect to the direction of the traveling magnetic field. This type of case occurs if one of the primaries mounted under the carriages of a train, driven by linear induction motors, is twisted with respect to the secondary (see Fig. 5-1). The following analysis concerns this problem.

- **Computational model**

The problem discussed here does not differ much from that analyzed in the preceding chapter, linked with the X-Y motor or motor with the spherical rotor. It means that the general mathematical model derived in chapter 3 can be directly applied here. To relate the linear motor with twisted primary to the above-mentioned model, first it has to be placed in an appropriate coordinate system.

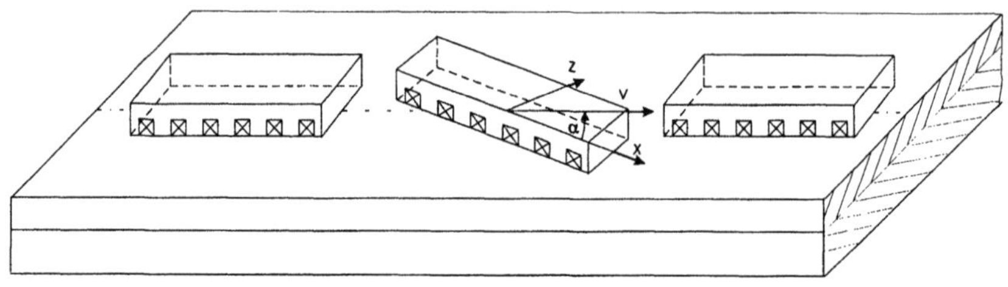

Figure 5-1. Illustration of the primary twist with respect to the direction of motion of the secondary of a flat linear induction motor

When the primary is twisted with respect to the secondary motion, then – neglecting the finite length and width effects – the speed responsible for the induction phenomena is the secondary speed component v_z, parallel to the direction of the field motion (Fig. 5-2). It gives along with the speed of the secondary the following relation:

$$v_x = v \cos \alpha \tag{5.1}$$

The force F_x, being a function of v_x, acts on the secondary in the x direction. The force F_v, which contributes to the secondary movement, is smaller than F_x and equal to:

$$F_v = F_x \cos \alpha \tag{5.2}$$

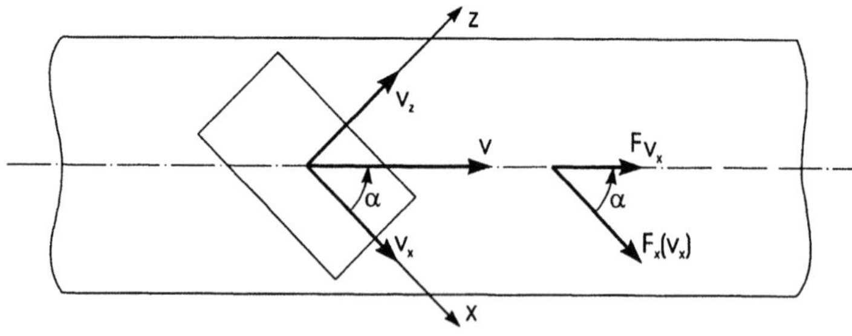

Figure 5-2. Illustration to Eqns. 5.1 and 5.2

The motor performance can now be determined using the multi-harmonic model defined in chapter 3.

-   Motor performance

As an example for modeling of a twisted primary, the flat linear motor produced in Poland has been chosen. Its constructional data are as follows:

Primary length	$L = 0.21$ m
Primary width	$W = 0.1$ m
Pole pitch	$\tau_z = 0.0501$ m
Number of slot per pole per phase	$q = 1$
Slot opening	$b_1 = 8$ mm
Number of wire per slot	$w = 210$
Tooth pitch	$\tau_t = 16.7$ mm
Primary linear current density	$J_{mzv} = \dfrac{6\sqrt{2}Iw}{\pi b_1 v}\sin\left(v\dfrac{\pi q b_1}{2\tau_x}\right)$
Air gap	$g = 1.5$ mm
Secondary aluminum plate thickness	$d = 5$ mm
Conductivity of aluminum plate	$\gamma_{Al} = 38.2{\cdot}10^6$ S/m
Conductivity of back iron	$\gamma_{Fe} = 5.91{\cdot}10^6$ S/m

The values of other quantities used in the calculations are as follows: $\mu_{Fe} = 1000{\cdot}\mu_o$, $\tau_{sx} = 0.42$ m, $\tau_{sz} = 0.4$ m. The particular values of τ_{sx} and τ_{sz} were chosen to insure no interaction between adjacent primaries in the computational model [37]. The computations were made at constant current $I = 3$ A and constant supply frequency $f = 50$ Hz.

The calculation results are shown in Figs. 5-3 – 5-5. in the form of magnetic field and force density distributions on the secondary surface. They were determined at a secondary speed 4.5 m/s for two different primary positions: $\alpha = 0$ and $\alpha = \pi/3$ rad. The synchronous speed was 5 m/s at 50 Hz.

The distributions of the normal component of magnetic flux density are shown in Fig. 5-3. Figs. 5-4 and 5-5 show the distributions of tangent force $\vec{f}_t = \vec{f}_x + \vec{f}_z$ and the electrodynamic force f_y normal to the secondary. When the primary is twisted some deformation with respect to the z-axis can be noticed. It is caused by the primary edges, thus it will not occur for an infinitely long and wide motor. The tangential forces act on the primary as they act to turn it into the normal position, parallel to the secondary motion. The levitation force f_y, which is particularly high at the primary transverse edges, is not the same on both sides of the z-axis at primary twist. Since it is higher on the exit edge, it tends to tilt the primary. Of cause, the normal magnetic attractive force also affects the primary, and the primary behavior will depend on the resultant effect of these two forces.

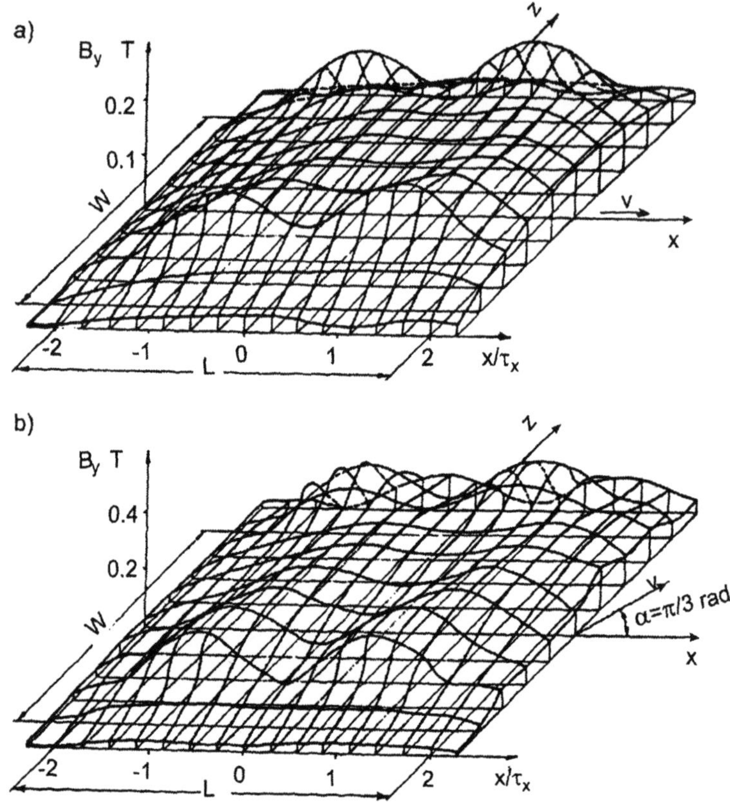

Figure 5-3. Normal component of magnetic flux density distribution on the secondary surface at two primary positions: (a) $\alpha = 0$, (b) $\alpha = \pi/3$ *rad.*

The influence of primary twist on motor mechanical characteristics is illustrated in Figs. 5-6 – 5-9. Fig. 5-6 shows the share of the twisted motor in driving the vehicle the motor is attached to. This contribution sharply declines as the angle α increases. It is caused not by the transverse edge effects, which contributes to the rise of braking transverse force component F_x, but by a decrease of the driving force component $F_v = F_x \cos\alpha$ (see Fig. 5-2). The characteristics of mechanical power and secondary efficiency vs. twist angle α, calculated at different speeds are shown in Figs. 5-7 and 5-8.

The mechanical power was calculated as follows:

$$P_m = vF_v \tag{5.3}$$

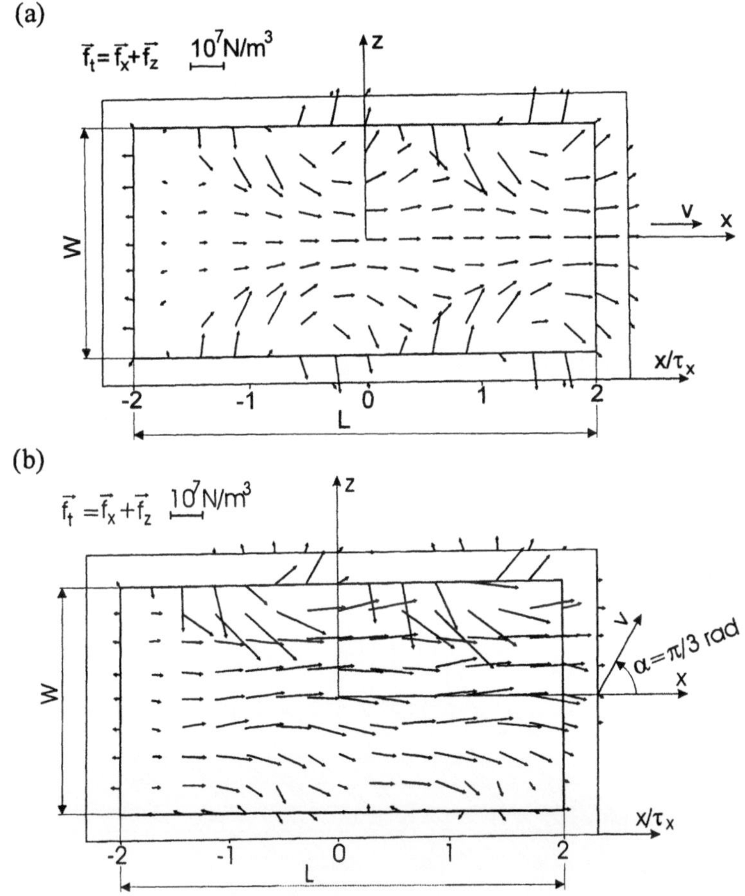

Figure 5-4.　Tangential force density distribution on the secondary surface at two primary positions: (a) $\alpha = 0$, (b) $\alpha = \pi/3$ rad.

and the efficiency as the ratio:

$$\eta = \frac{P_m}{P_r} \tag{5.4}$$

where P_r is the power crossing the air gap.

The two quantities P_m and η are referred to their values at $\alpha = 0$ and are compared with $\dfrac{Q_\alpha}{Q_{\alpha=0}} = \cos\alpha$ curve (Q – quantity). It enables one to see the influence of transverse edge effects because their lack (for an infinitely wide motor) makes these characteristics equal to $\cos\alpha$

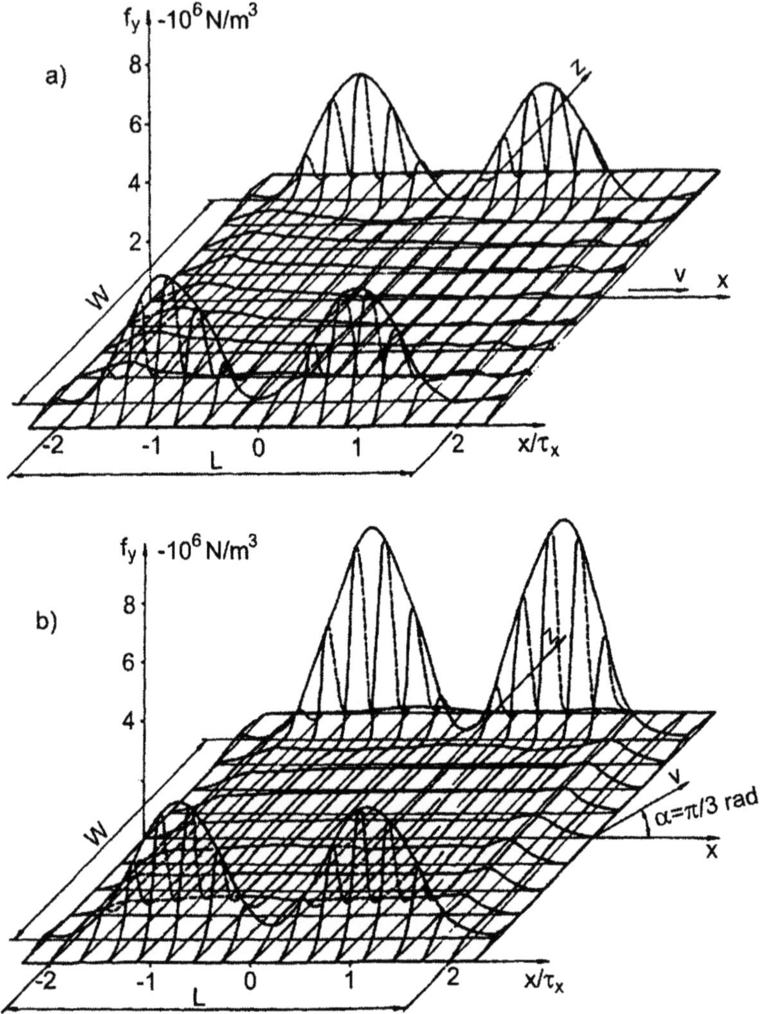

Figure 5-5. Normal component of force density distribution on the secondary surface at two primary positions: (a) $\alpha = 0$, (b) $\alpha = \pi/3$ rad

Fig. 5-9 shows the resultant force acting on the secondary, which is drawn against the secondary speed at different primary twists. This force is a geometrical sum of the F_z and F_x components:

$$F = \sqrt{F_x^2 + F_z^2}$$

(5.5)

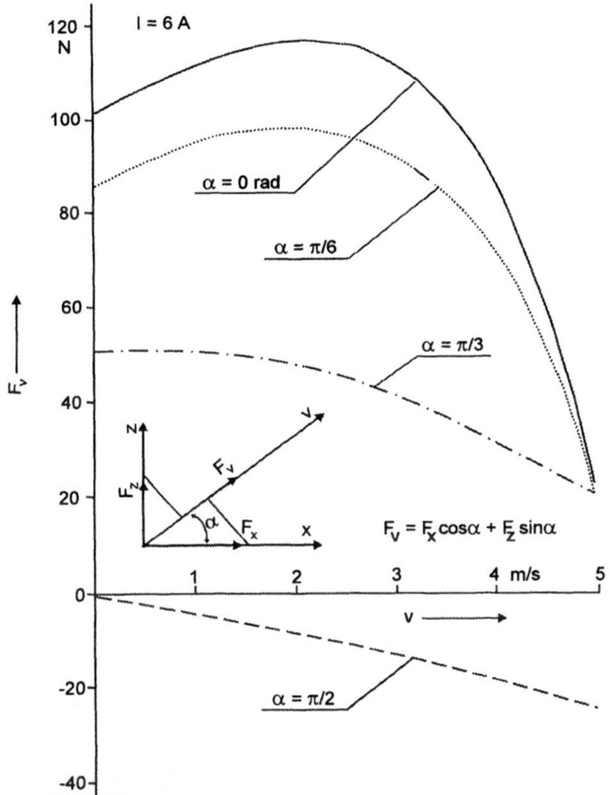

Figure 5-6. Characteristics of force acting in the direction of secondary movement

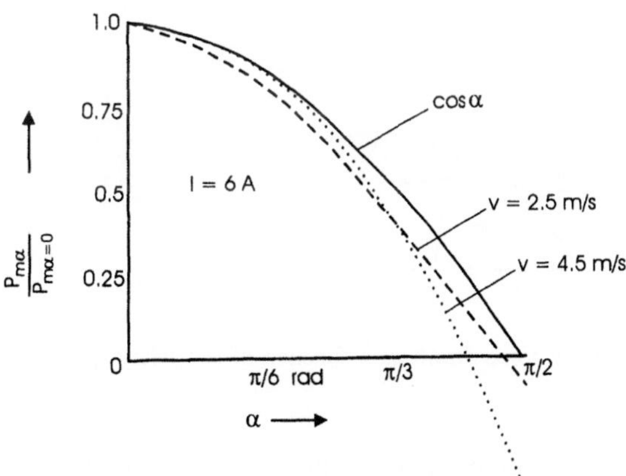

Figure 5-7. Influence of primary twist on mechanical power

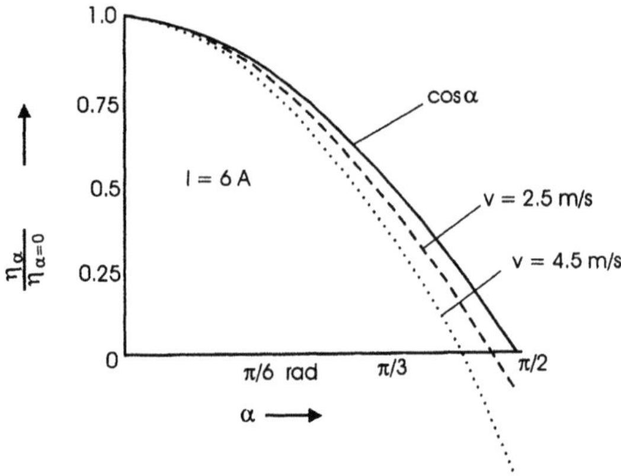

Figure 5-8. Impact of primary twist on secondary efficiency

$$F_t = f(v) \quad \underset{\text{100 N}}{\longmapsto} \quad F = \sqrt{F_x^2 + F_z^2}$$

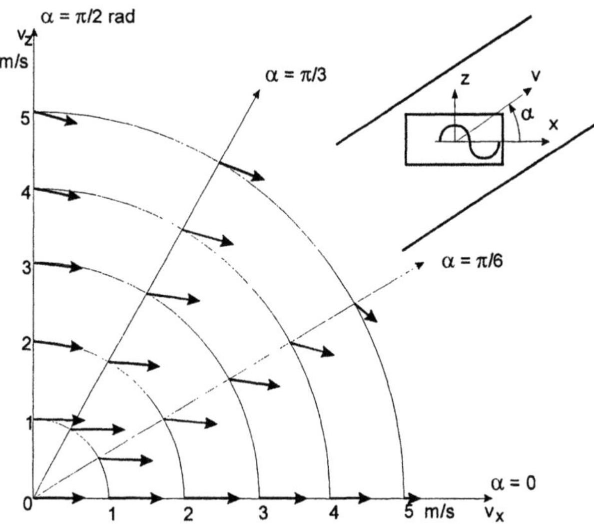

Figure 5-9. Tangential force drawn as a function of secondary speed at different primary twists

The direction of force changes slightly when the primary twist occurs. This is caused by the transverse edge effects, which occur when the lateral secondary speed component v_z exists.

To partially verify the computational model measurements of magnetic flux density distribution were carried out at zero speed. The results are shown in Fig. 5-10 and are compared with the ones obtained from simulation.

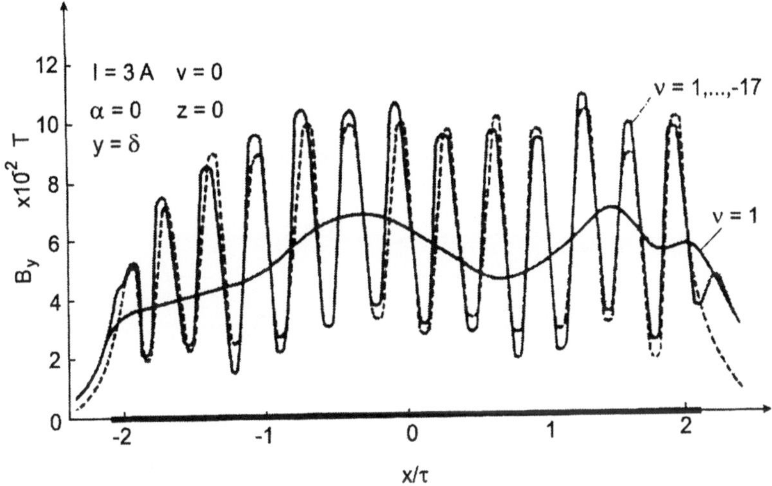

Figure 5-10.   Measured (------) and computed (———) magnetic flux density distributions on the secondary surface along the *x* axis

More examples of an application of the mathematical model derived for motors with two degrees of mechanical freedom for analysis of linear and rotary cylindrical motors are presented in [17, 38, 39, 42].

In general, there are number of publications on linear motors. Among them are books [28, 29, 30, 52, 58, 63, 67] and papers [6, 7, 14, 31, 53, 56, 57].

2.   DISC-TYPE INDUCTION MOTORS

There are a few versions of disc induction motors and they can be classified into two major groups:
- single-sided motors and
- double-sided motors.

Two motors, which are representatives of these groups, are discussed further in this section. Both motors were designed to drive water pumps and their rotors were integrated with the pump turbines operating as aggregates with wet rotors. The stators have two-phase windings supplied from a single-phase source. Therefore the motors represent the group of induction machines with unsymmetrical winding supplied by nonsymmetrical currents.

2.1 Single-sided disc induction motor

The diagram of the single-sided disc motor structure is shown in Fig. 5-11.a. The armature (stator) consists of toroidal laminated iron with radial slots, and two windings: the main winding U1 – U2 and the auxiliary winding Z1 – Z2 (Fig. 5-11.b), which is connected in series with a capacitor. The rotor has a double-layer structure made of a steel disc, covered with a copper sheet. The stator core is separated from the water (which the rotor operates in) by a layer made of nonmagnetic material.

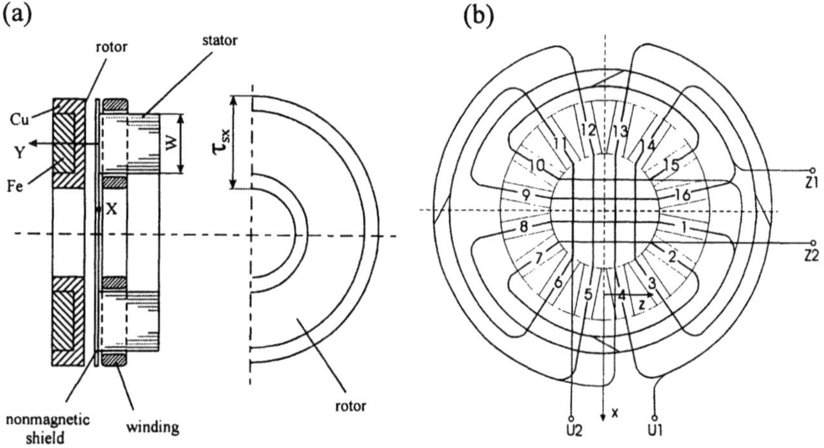

Figure 5-11. a) Diagram of a single-sided disc motor, b) winding distribution: *U1-U2* main winding, *Z1-Z2* auxiliary winding

- **Computational model of the motor**

The motor should be analyzed in a cylindrical coordinate system. However, if the motor inner diameter is relatively large the analysis can be carried out in a perpendicular coordinate system, where the equations are much simpler. The calculation model of the motor used here is defined in Cartesian's coordinate system and refers to the infinitely long flat linear motor with finite stator and rotor widths (Fig. 5-12). The mathematical model is therefore a particular case of the more general model defined for motors with two degrees of mechanical freedom in chapter 3.

To more precisely describe the model currently discussed it is enough to formulate the function that describes the stator current density. If the slot harmonics h are taken into account the current density of pth phase is defined by the following Fourier's series:

$$J_x^p(t,z) = \sum_v \sum_h \sum_q \sum_t J_{mxvhqt}^p \exp\left[j\left(\omega t - \frac{\pi}{\tau_{zvh}} z \right) \right] \qquad (5.6)$$

where:

J_{mxvhqt}^p - amplitude of $vhqt^{th}$ harmonic,

$$\tau_{zvh} = \frac{1}{q\dfrac{2h}{\tau_t} + t\dfrac{v}{\tau_z}}, \qquad q, t = 1, -1, \qquad (5.7)$$

τ_t – tooth pitch,
τ_z – pole pitch,
$v - 1,3,5,7,...$ – phase harmonics,
$h - 0,1,2,3,...$ – tooth harmonics.

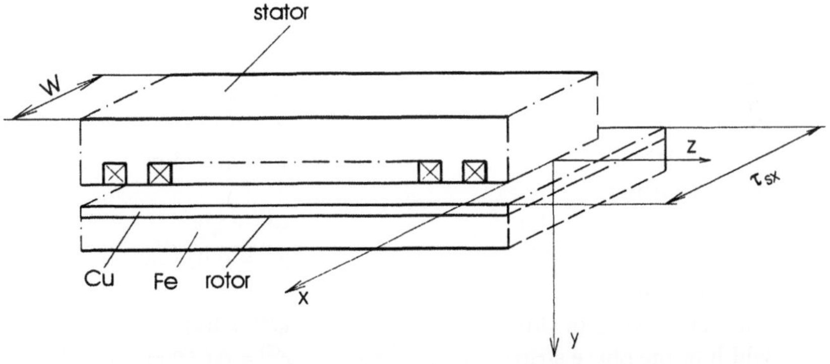

Figure 5-12. Flat model of disc motor

To take into account the finite width of the stator the appropriate harmonics of the current density should be considered. Referring to the general model in chapter 3 the current density of the pth phase is described by the function:

$$J_x^p(t,z) = \sum_k \sum_u \sum_v \sum_h \sum_q \sum_t \frac{1}{2} J_{mxkvhqt}^p \exp\left[j\left(\omega t - \frac{\pi}{\tau_{xku}} x - \frac{\pi}{\tau_{zvh}} z \right) \right] \quad (5.8)$$

where

$$J_{mxkvhqt}^p = \frac{4}{\pi k} \sin\left(k \frac{W\pi}{2\tau_{sx}} \right) J_{mxvhqt}^p \qquad (5.9)$$

$$k = 1,3,5,...$$

$$\tau_{xku} = \frac{\tau_{sx}}{uk},\qquad\qquad(5.10)$$

$u = 1,-1$
W – armature width,
τ_{sx} – rotor width.

The resultant current density of the stator is a sum of current densities of two phases.

- **Motor characteristics**

The single-sided disc motor that is considered was designed and built by BOBRME, a Company in Katowice, Poland. The motor's data are as follows:

Armature

stator width	$W = 0.0316$ m
pole-pitch	$\tau_z = 0.116$ m
trooth-pitch	$\tau_t = 0.0073$ m

Main winding:

number of wires in slot	$w^{(1)} = 200$
width of the phase strip	$c^{(1)} = 0.058$ m

Auxiliary winding:

number of wires in slot	$w^{(2)} = 300$
width of the phase strip	$c^{(2)} = 0.058$ m

Air gap $g_r = 3.5$ mm

Rotor:

width	$\tau_{sx} = 0.05$ m
thickness of copper layer	$d_{Cu} = 1$ mm
copper conductivity	$\gamma_{Cu} = 46.8{\cdot}10^6$ S/m
back iron conductivity	$\gamma_{Fe} = 4.56{\cdot}10^6$ S/m

Any shield on the stator surface was neglected in the calculations. In order to relate the calculations to the real object the measurements of the stator currents were carried out at the frequency supply $f = 50$ Hz and the rotor slip $s = 0.2$. The phase angle between the currents was equal $\psi = 0.803$ *rad*. The calculation results are shown in Figs. 5-13 and 5-14 as the magnetic flux density and the corresponding tangential force component distribution on the rotor surface over one pole pitch.

The nonuniform distribution of magnetic field results from the asymmetry of the stator currents. This contributes to nonuniform distribution of forces, which act in some area of the rotor in the opposite direction to the rotor motion.

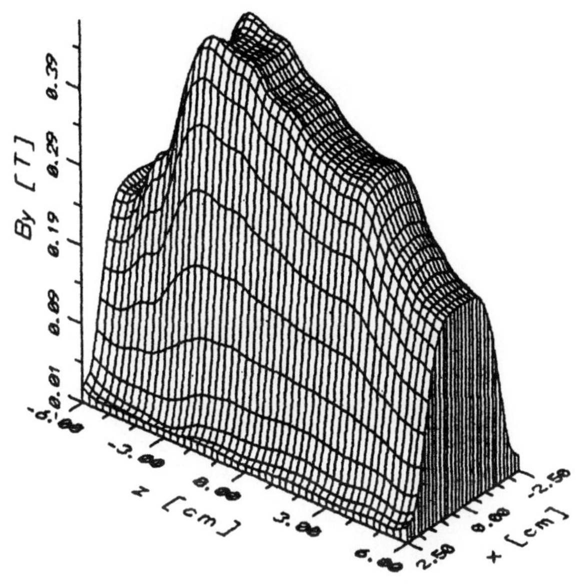

Figure 5-13. Magnetic flux density distribution on the rotor surface

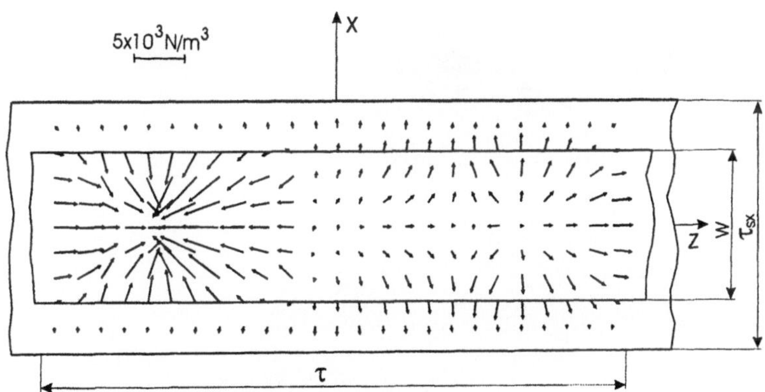

Figure 5-14. Tangential force component distribution on the rotor surface

In order to partially verify the mathematical model measurements of magnetic flux density were carried out in the middle of stator ring, *5 mm* from its surface without the presence of the rotor. The measured currents were: I_1 = 1.73 A, I_2 = 0.93 A, ψ = *0.0377 rad.* Both, experimental and test results are shown in Fig. 5-15. The major discrepancies between the characteristics are in the range of small magnetic flux densities, which occur on the edge of both windings. There, the additional magnetic field, produced by the currents flowing in the winding ends occurs, which was not considered in the computational model and the share of which is particularly high if there is no rotor.

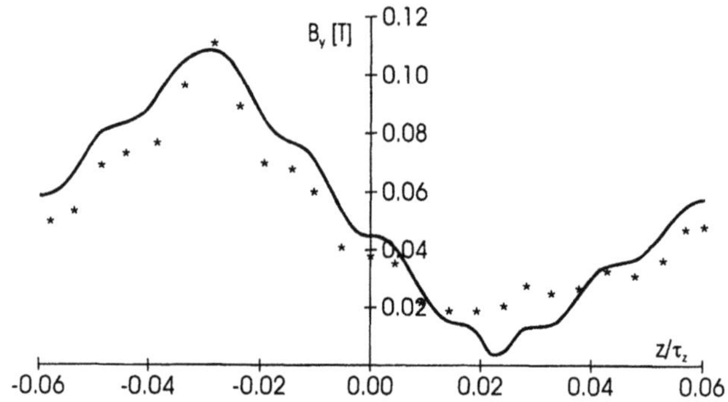

Figure 5-15.   Computational and experimental results of magnetic flux density distribution on the stator circumference

To qualitatively compare the space distribution of the magnetic field the simulation and test results are shown in Fig. 5-16.a and b.

(a)

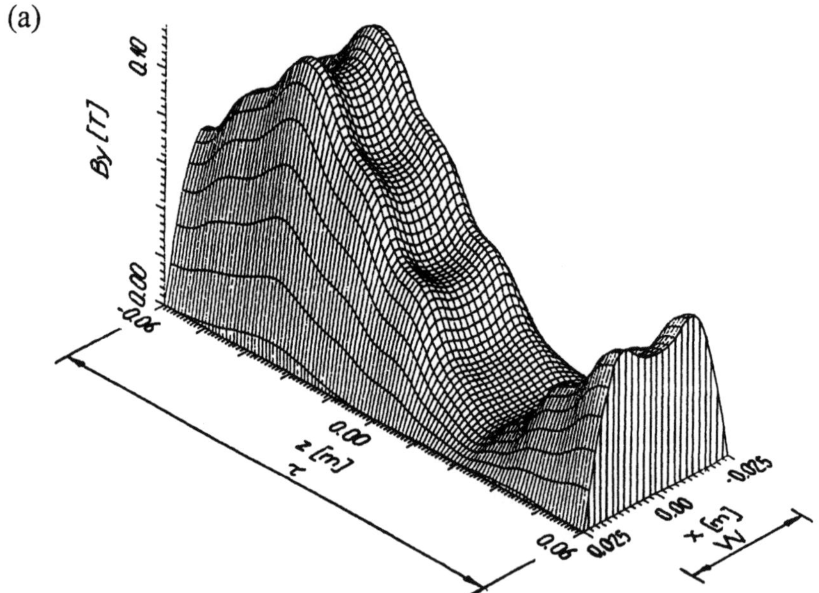

(b)

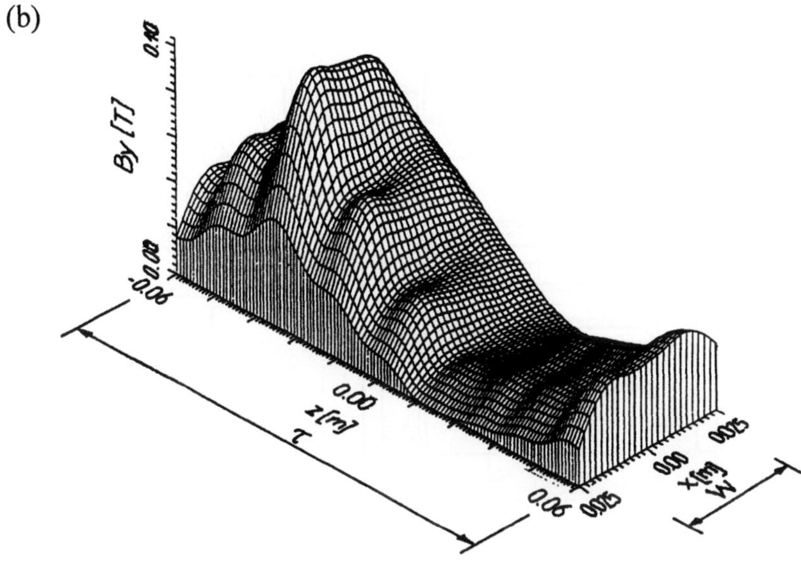

Figure 5-16. Characteristics of magnetic flux density distribution: (a) theoretical and (b) experimental characteristic

The test results are related to the flat computational model, and the shape of the two characteristics is similar. The influence of stator teeth is visible, however it does not occur in the experimental characteristic on the inner part of rotor disc (negative values on x axis) due to saturation of the more narrow part of stator teeth.

More details concerning a simulation of this type motor can be found in [47].

2.2   Double-sided disc induction motor

The motor that is considered in this section has a double-sided stator and twin rotors (Fig. 5-17). A stator with radial slots is placed between two double-layer rotor discs. The stator winding is the Gramme's type, made of separate coils wound on the stator core. These coils can be connected in different systems forming a three or two phase winding of two or more magnetic poles. The slotless version of this type of the disc motor is called a torus motor in the literature [4].

To analyze the electromagnetic field and motor performance a computational model similar to that used for the single-sided disc motor and flat linear motor with the infinite long stator and rotor is applied (Fig. 5-12).

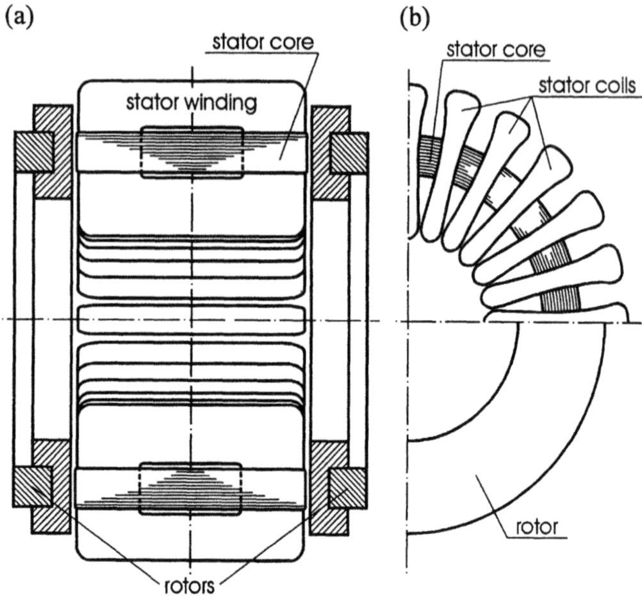

Figure 5-17. Scheme of a disc motor with double-sided stator and twin rotors

Since only one side of the motor is considered in this model, to calculate the torque and power of the entire motor, the results obtained for one side should be multiplied by 2.

The motor prototype, which is the subject of analysis, is shown in Fig. 5-18. It was manufactured by AQUA-SHUT Company. Its parameters are as follows:

medium diameter of the motor disc	$D_{av} = 81.96$ mm
armature ring width	$W = 10$ mm
pole pitch	$\tau_z = 64{,}2$ mm
number of magnetic poles	$2p = 4$
number of slots per pole per phase	$q = 3$
number of turns in the coil: phase 1	$T_{1k} = 320$
phase 2	$T_{2l} = 320$
resistances of two phases	$R_1 = 73.4$ Ω
	$R_2 = 73.4$ Ω
leakage reactances	$X_{s1} = 82$ Ω
	$X_{s2} = 82$ Ω
air-gap	$g = 0.8$ mm
width of the rotor ring	$\tau_{sx} = 15$ mm
thickness of the copper layer	$d = 1$ mm

The electromagnetic field of the motor and forces acting on the rotor calculated when the motor was supplied from a constant current source are presented in [12]. The distributions of these quantities on the rotor surface are similar to those of the single-sided motor shown in the previous section.

Figure 5-18. Prototype of a double-sided disc motor

In this section the motor performance is considered when the two-phase winding is supplied from a single-phase and two-phase constant voltage source. The circuit diagram of the motor is shown in Fig. 5-19. The voltage equations that correspond with this circuit are as follows:

$$V_1 = E_1 + (R_1 + jX_{s1})I_1 \tag{5.11}$$

$$V_2 = E_2 + [R_2 + j(X_{s2} - X_c)]I_2 \tag{5.12}$$

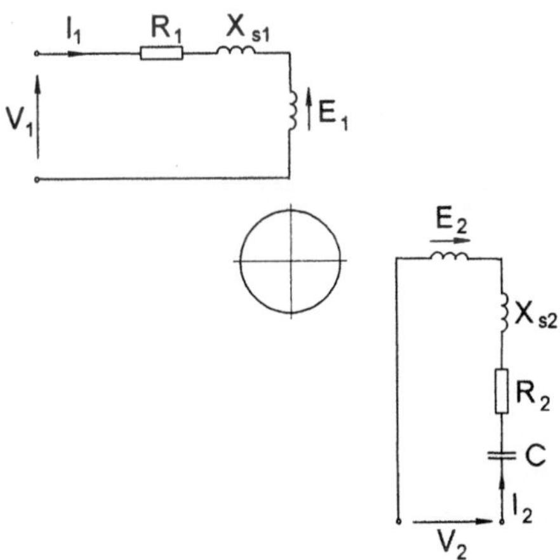

Figure 5-19. Circuit diagram of the disc motor

The circuit parameters R_1, R_2, X_{s1} and X_{s2} are resistances and leakage reactances of the two windings. These can be determined using conventional methods. E_1 and E_2 are the voltages induced in the windings by the magnetic fluxes linked with them. These magnetic fluxes are generated by the two windings and are the sums of all harmonics expressed in the field model. The multi-harmonic circuit model of the motor is presented in chapter 3. According to this model, the equivalent circuit of both phases is as shown in Fig. 5-20. The EMFs E_1 and E_2 are now shown as the voltage drops across the "self impedances" $Z_{11kvhqt}$ and $Z_{22kvhqt}$ and across "mutual impedances" $Z_{12kvhqt}$ and $Z_{21kvhqt}$, where the voltages of the $kvhqt^{th}$ harmonics depend on the currents flowing in the other windings.

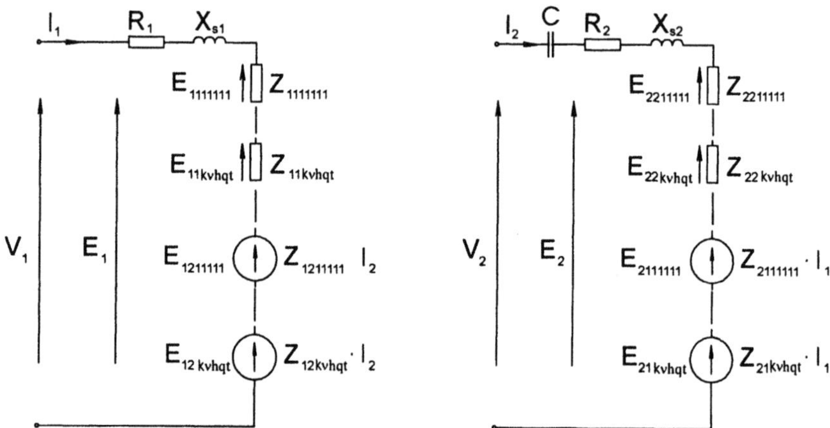

Figure 5-20. Equivalent circuit of the disc motor with higher harmonics

The $kvhqt^{th}$ harmonics of impedances might be rolled up into the resultant impedances expressed in the following way:

$$Z_{11(22)} = \sum_k \sum_u \sum_v \sum_h \sum_q \sum_t Z_{11(22)kvhqt} \qquad (5.13)$$

$$Z_{12(21)} = \sum_k \sum_u \sum_v \sum_h \sum_q \sum_t Z_{12(21)kvhqt} \qquad (5.14)$$

According to this the equivalent circuit, which was used to farther calculations, is shown in Fig. 5-21. It must be pointed out that the values of the impedances depend on the rotor speed, so they have to be determined for the different speeds over the whole range the motor is to be operated.

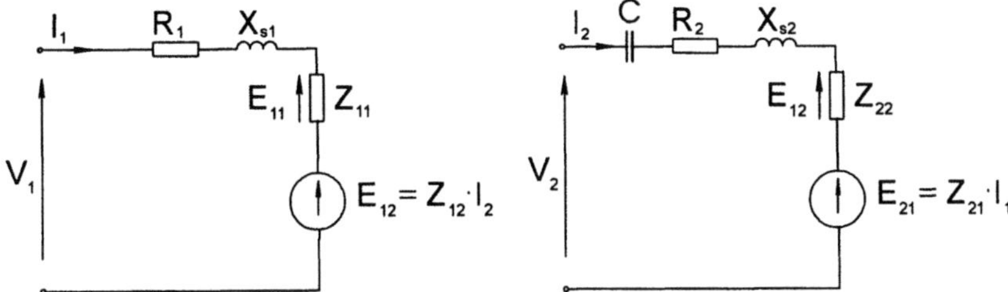

Figure 5-21. Equivalent circuit with the resultant impedances

The voltage equations corresponding to the equivalent circuit shown in Fig. 5-21 are as follows:

$$V_1 = (R_1 + jX_{s1})I_1 + Z_{11} I_1 + Z_{12} I_2$$
$$V_2 = [R_2 + j(X_{s2} - X_c)]I_2 + Z_{22} I_2 + Z_{21} I_1 \qquad (5.15)$$

where:

$$Z_{11} = R_{11} + jX_{11}, \qquad Z_{12} = R_{12} + jX_{12},$$

$$\qquad (5.16)$$

$$Z_{22} = R_{22} + jX_{22}, \qquad Z_{21} = R_{21} + jX_{21}.$$

These equations enable the determination of the currents at the given supply voltages. Once the currents have been determined their values have to be used to calculate the forces and powers from the expressions derived in the field model (see chapter 3).

The circuit parameters R and X_s were found using the classical approach [49]. The other parameters of the equivalent circuit shown in Fig. 5-20, were calculated from the equations 5.13 and 5.14. As was mentioned above, their values vary with the rotor speed and this is shown in Fig. 5-22.

On the basis of the equivalent circuit the stator currents were determined when the motor was supplied from an ac. voltage source of 220 V and a frequency of *50 Hz*. The calculations were carried out first for the motor supplied from a two-phase symmetrical source, then from the single-phase voltage source. In the latter case, the capacitor of *13 μF* capacitance was connected in the auxiliary phase. The characteristics of the currents versus speed for the single-phase supply are shown in Fig. 5-23. The theoretical characteristics are compared with the ones obtained from the test carried out on the real object.

As one can see, the currents do not change much due to the relatively large air gap and rotor cooper layer. The discrepancies between theory and

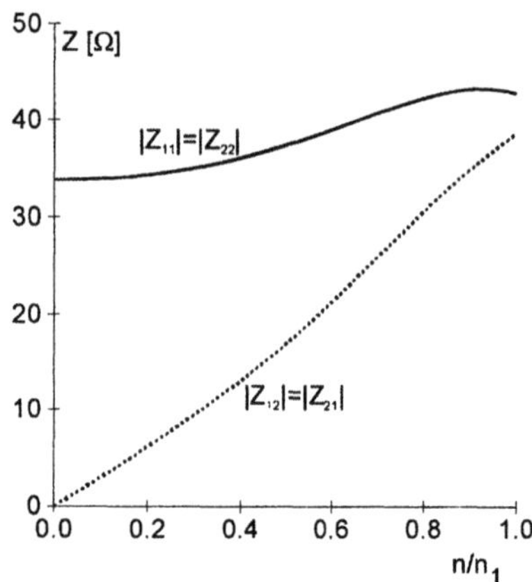

Figure 5-22. Variation of the resultant impedances with rotor speed

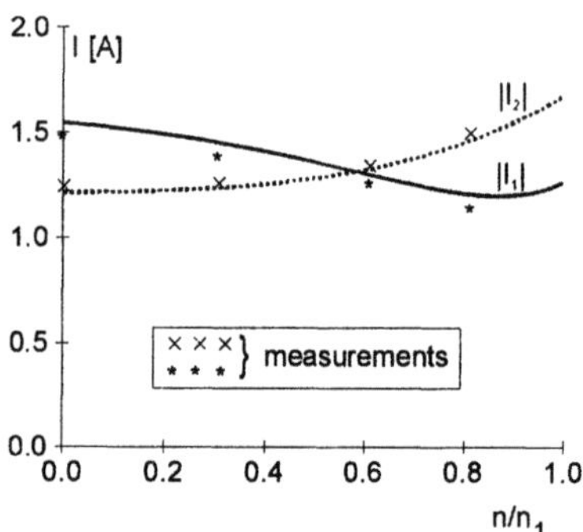

Figure 5-23.   Currents vs. speed characteristics calculated and measured using a single-phase supply of *220 V, 50 Hz* with *13 µF* capacitor

test are not significant. This indicates that the mathematical model describes the motor with a relatively good accuracy.

Once the currents were determined from the circuit model they allowed the calculation of the other quantities such as torque, power and efficiency using the formula derived from the field model. Thus the torque developed by the motor can be found from the expression:

$$T = \frac{D_{av}}{2} \sum_k \sum_u \sum_v \sum_h \sum_q \sum_t \frac{\Delta P_{kvhqt}}{v_{vhqt}\, s_{vhqt}} \qquad (5.17)$$

where:

$$v_{vhqt} = 2 \cdot \tau_{zvhqt} \cdot f \tag{5.18}$$

is the speed of the $k\,vhqt^{th}$ field harmonic.

The power:

$$\Delta P_{k\,vhqt} = \mathrm{Re}\left\{ \int_{-\frac{W}{2}}^{\frac{W}{2}} \int_{0}^{\pi D_{av}} S_{ykvhqt} \cdot dx \cdot dz \right\} \tag{5.19}$$

is the power loss in the rotor. It is expressed in terms of the Poynting's vector S_y, which is integrated over the whole rotor surface. The Poynting's vector for the $k\,vhqt^{th}$ harmonic is defined in the following way

$$S_{ykvhqt} = 2 \cdot \frac{1}{2} \left(E_{mzkvhqt} \cdot H^{*}_{mxkvhqt} - E_{mxkvhqt} \cdot H^{*}_{mzkvhqt} \right) \tag{5.20}$$

and it represents the rotor power loss surface density when the coordinate system, in which the field analysis was carried out, is attached to the rotor as it was done here.

The electromechanical characteristics such as: torque, output power and power angle versus relative rotor speed calculated for two-phase and single-phase supplies are shown in Figs. 5-24 and 5-25. They are compared with the ones obtained from the measurements.

More data on the performance of this type of motor can be found in [12, 13, 49].

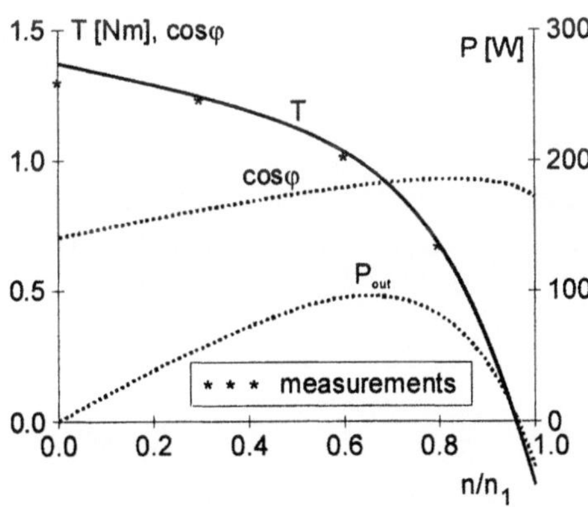

Figure 5-24. Mechanical characteristics calculated and measured at a single-phase supply of *220 V, 50 Hz* with *13 μF* capacitor

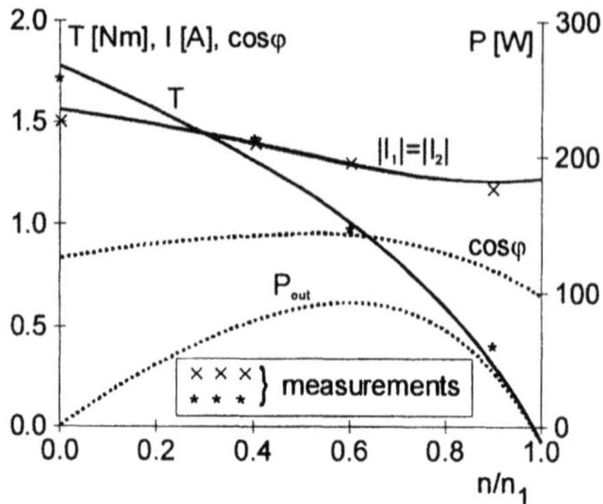

Figure 5-25. Electromechanical characteristics calculated and measured at a two-phase symmetrical supply of *220 V, 50 Hz*

3. MHD MACHINE – INDUCTION PUMP

Three-phase inductors with a magnetic traveling field, which serve as a primary of linear induction motors, are often used to pump a liquid metal or to stir it in induction stirrers [9]. Due to the finite length of inductor an alternating magnetic field is generated as an additional component to the magnetic traveling field. When an induction pump is used with a rectangular transport channel and a flat inductor (see Fig. 5-26), this component deforms a force distribution in liquid metal and causes some local whirls [18, 48]. This may play a positive role in the stirring of molten metal but it cannot be tolerated in the channel of an induction pump, where a smooth flow is required. In that case the alternating magnetic field should be eliminated. This can be done by using an additional compensating winding or by

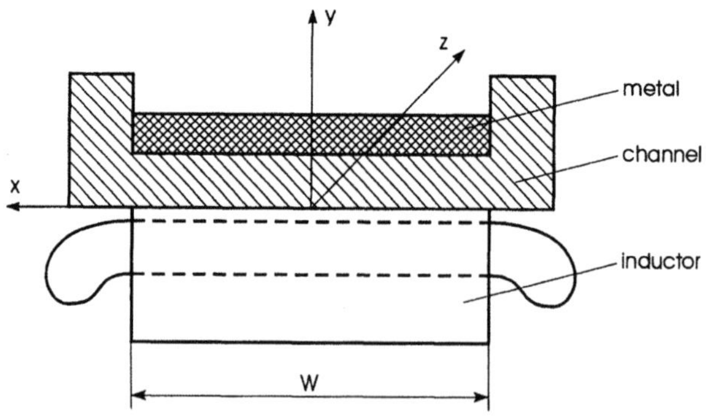

Figure 5-26. Cross-section of an induction pump

applying a special type of three-phase winding that eliminates or reduces the alternating field component [18]. This type of winding is shown in Fig. 5-27. It is a three-phase, two-layer, two-pole winding with the external sides of the coils placed out of the laminated iron core.

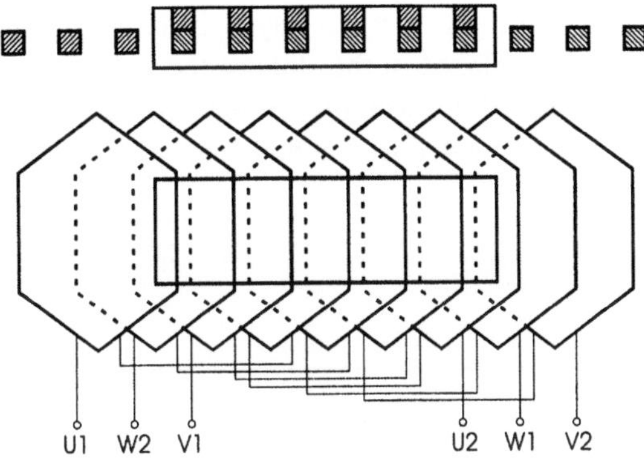

Figure 5-27. Three-phase, two-layer, two-pole winding of an inductor

To examine the magnetic field and force distribution in molten metal of an induction pump without and with a compensating winding, the mathematical model derived for motors with two degrees of mechanical freedom will be used with some modification.

- **Mathematical model**

The velocity of the molten metal in the pump depends on the electromagnetic field as well as on the shape of the channel. If the liquid metal is driven by the traveling field of an inductor placed below a rectangular channel, the velocity is non-uniform across the channel as shown in Fig. 5-28.a. It is very difficult to determine the electromagnetic field with such a distribution using an analytical method. For that reason an assumption

(a)   (b)

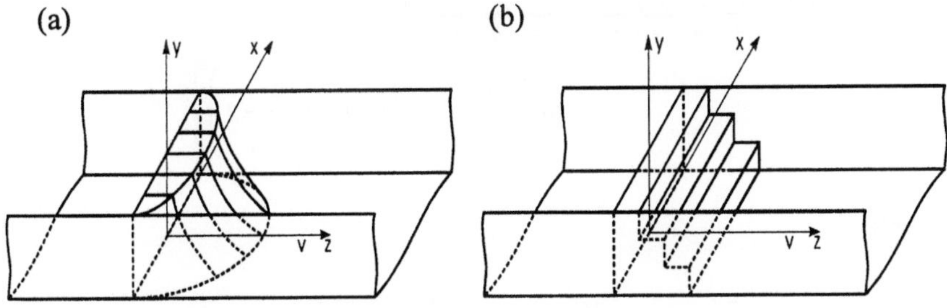

Figure 5-28.   Distribution of liquid metal velocity in the channel: (a) real distribution, (b) assumed distribution

has been made that the liquid metal is represented by a number of layers with different but constant velocity in each layer as shown in Fig. 5-28.b.

The computation model of the pump used for the electromagnetic field description is the N-layer structure shown in Fig. 5-29. It is, in general, similar to the one defined in chapter 3. The only difference is in the current layer description. Here, the current layer is represented by the current density of the winding placed in the slots of the inductor (one part) and the density of another part of the winding, called here a compensating winding, placed outside of the laminated iron at both ends of the inductor.

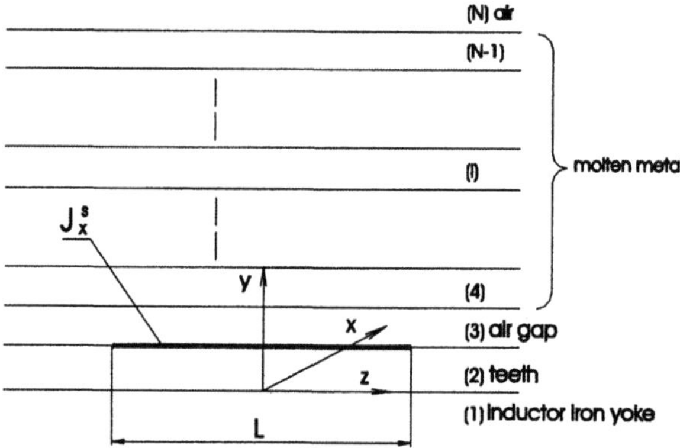

Figure 5-29. Computational model of an induction pump

The functions that describe the current densities of both parts of the winding are shown in Figs. 5-30.a and b. Both functions can be expressed by Fourier's series. The first one, describing the current density of the main part of the winding has the following form (see chapter 3):

$$J_x^s(t,z) = \sum_l \sum_k \sum_u \sum_r \sum_v \sum_h \sum_q \sum_t \frac{1}{4} J_{mxvhqt}^s$$

$$\cdot \exp\left[j\left(\omega t - \frac{\pi}{\tau_{xku}} x - \frac{\pi}{\tau_{zlrvh}} z \right) \right] \tag{5.21}$$

where

$$J_{mxklvhqt}^s = \frac{16}{\pi^2 kl} \sin\left(k \frac{W\pi}{2\tau_{sx}} \right) \sin\left(l \frac{L\pi}{2\tau_{sz}} \right) J_{mxvhqt}^s \tag{5.22}$$

$$\tau_{xku} = \frac{\tau_{sx}}{uk}, \qquad \tau_{zhvlr} = \frac{\tau_{zhv}\tau_{sz}}{lr\tau_{zhv} + \tau_{sz}} \tag{5.23}$$

$h = 0, 1, 2, 3, \ldots$ tooth harmonics

$v = 1, 3, 5, \ldots$ phase harmonics

$k, l = 1, 3, 5, \ldots$ harmonics generated by the finite width W and length L
        of the inductor current layer

$u, r, q, t = 1, -1$

The Fourier's series that represents the function's current density of the compensating winding may be derived in a similar way. The current density of the pth phase is given by (Fig. 30-b)

$$J_x^{c(P)}(t, z) = \sum_{\xi} J_o^{c(P)} \exp\left[j\left(\omega t + \varphi^{c(P)}\right)\right]\frac{4}{\pi\xi}$$
$$\sin\left(\xi\frac{c^c\pi}{2\tau^c}\right)\cos\left[\xi\left(\frac{\pi}{\tau^c}z + \psi^c\right)\right] \tag{5.24}$$

where:

$$J_o^{c(P)} = \frac{\sqrt{2}I^{c(P)}w^{c(P)}}{c^c}$$

$w^{c(p)}$ — number of wires in one coil of the compensating winding of the
      p^{th} phase,

c^c — coil thicness,

$I^{c(p)}$ — rms current of the pth phase in c compensating winding.

(a)   (b)

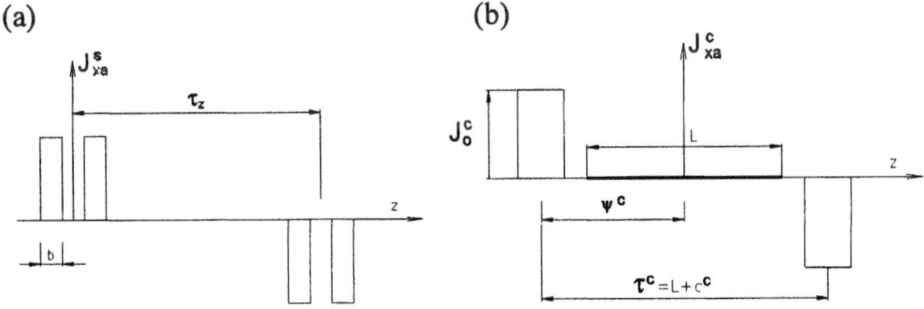

Figure 5-30.   Function of current density of phase A: (a) for the main part of the winding contained in the slots, (b) for the compensating part of the winding placed off the inductor core

Function 5.24 written in another form is as follows:

$$J_x^{c(p)}(t,z) = \sum_o \sum_\xi J_{m\xi o}^{c(p)} \exp\left[j\left(\omega t + \frac{\pi}{\tau_\xi^c} z \right) \right] \tag{5.25}$$

where:

$$J_{m\xi o}^{c(p)}(t,z) = \frac{1}{2} J_o^{c(p)} \frac{4}{\pi\xi} \sin\left(\xi \frac{c^c \pi}{2\tau^c} \right) \exp\left[j\left(\varphi^{c(p)} - \xi o \psi^{c(p)} \right) \right] \tag{5.26}$$

$$\xi = -1, 3, 5, \ldots, \qquad o = 1, -1,$$

$$\tau_\xi^c = \frac{1}{\xi} \tau^c .$$

The resultant current density of m phases is given by:

$$J_x^c(t,z) = \sum_o \sum_\xi J_{mx\xi o}^c \exp\left[j\left(\omega t + \frac{\pi}{\tau_\xi^c} z \right) \right] \tag{5.27}$$

where

$$J_{xm\xi o}^c = \sum_{p=1}^m J_{m\xi o}^c \tag{5.28}$$

If the finite length L^c and width W of the compensating winding are taken into account, the current density expressed by Fourier's series takes the following form:

$$J_x^c(t,z) = \sum_l \sum_k \sum_u \sum_r \sum_\xi \sum_o \frac{1}{4} J_{mxkl\xi o}^c \exp\left[j\left(\omega t - \frac{\pi}{\tau_{xku}} x - \frac{\pi}{\tau_{zlr\xi}} z \right) \right] \tag{5.29}$$

where

$$J_{mxkl\xi o}^c = \frac{16}{\pi^2 kl} \sin\left(k \frac{W\pi}{2\tau_{sx}} \right) \sin\left(l \frac{L^c \pi}{2\tau_{sz}} \right) J_{mx\xi o}^c \tag{5.30}$$

$$\tau_{xku} = \frac{\tau_{sx}}{uk}, \qquad \tau_{zlr\xi} = \frac{\tau_{\xi}^{c}\tau_{sz}}{lr\tau_{\xi}^{c} + \tau_{sz}} = \frac{\tau^{c}\tau_{sz}}{lr\tau_{\xi}^{c} + \xi\tau_{sz}} \qquad (5.31)$$

$k, l = 1, 3, 5, \ldots$
$u, r = 1, -1.$

The ith layer of molten metal moves in the z direction with speed v_{zi}. The slip s_i of the ith layer with respect to each harmonic of the magnetic field is given by equations

$$S_{zlrvhqt} = 1 - \frac{v_{zi}}{v_{zlvhr}} \qquad (5.32)$$

for the main winding, and

$$S_{zlr\xi} = 1 - \frac{v_{zi}}{v_{zlr\xi}} \qquad (5.33)$$

for the compensating winding, where

$$v_{zl\,vhr} = 2\tau_{zl\,vhr}\,f \quad \text{and} \quad v_{xlr\xi} = 2\tau_{zl\xi}\,f \qquad (5.34)$$

are the velocities of field harmonics of both windings.

The electromagnetic fields of both windings are determined next using the formulae derived in chapter 3.

- **Magnetic field and forces acting on molten metal**

The induction pump that is considered as an example of the MHD machine is the pump for molten zinc with the following characteristics:

Inductor length	$L = 0.6$ m
Inductor width	$W = 0.2$ m
Winding: 3-phase, 4-pole, two layers	
Pole pitch	$\tau_z = 0.15$ m
Tooth pitch	$\tau_t = 0.025$ m
Slot opening	$b_1 = 0.012$ mm
Number of slots per pole per phase	$q = 2$
Number of wires per slot	$w = 64$
Number of wires in the coil-side outside the laminated iron in one phase	$w^c = 64$

Distance between inductor and liquid metal	$g = 0.03$ m
Liquid metal thickness	$d = 0.1$ m
Channel width	$\tau_{sx} = 0.2$ m
Zinc conductivity	$\gamma_{Zn(500^{\circ}C)} = 2.7 \cdot 10^{6}$ S/m
Phase rms current of the inductor	$I = 27$ A
Current frequency	$f = 50$ Hz

The computational model used here for calculation consists of 7 layers, in which three represent a liquid metal with different speeds being in the range *1.5* to *9 m/s*. In general, the number of layers representing the liquid metal is not limited in the algorithm and for higher speeds this number should be increased.

The calculations of magnetic flux density, power losses and force density were carried out for the inductor with and without the compensating winding for $y = 0.03$ *m.*, i.e., at the bottom of the transport channel. The results obtained from calculations done for the inductor without compensation and with a compensating winding are shown in Figs. 5-31 – 5.33. Since there is symmetry in the distributions of all three quantities the characteristics are only drawn for one half of the inductor. From the magnetic flux density distribution one can see a substantial reduction of deformation after introducing a compensating winding. This can also be noticed in the tangential force distribution (Fig. 5-32). The local swirls at the inductor edges caused by the alternating magnetic field are significantly reduced

(a)

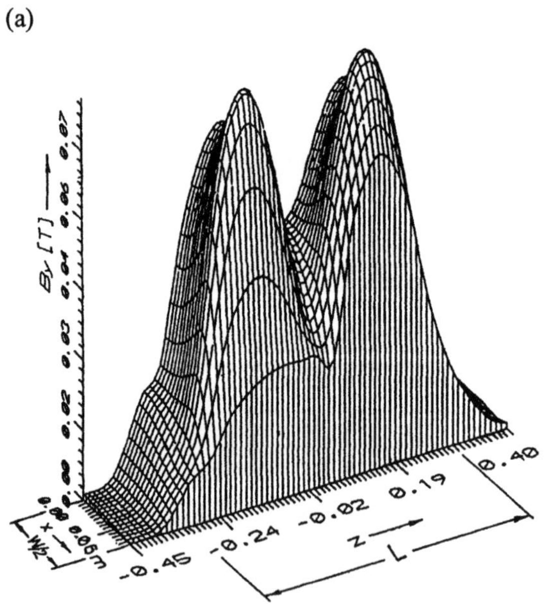

(b)

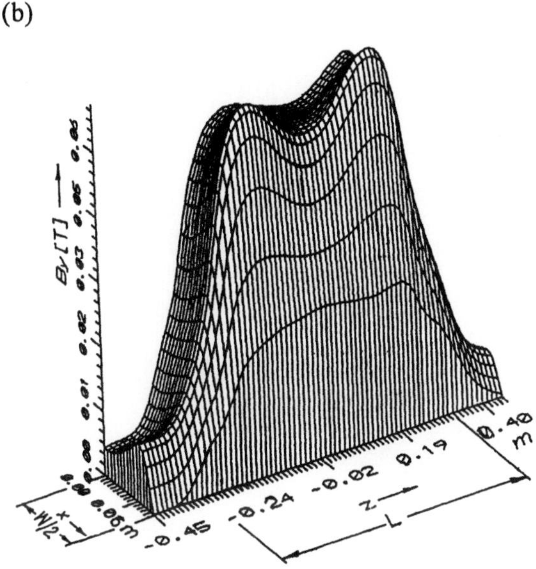

Figure 5-31. Magnetic flux density distribution at the channel bottom: (a) without compensating winding, (b) with compensating winding

when this field component is compensated. The compensation of alternating magnetic flux also contributes to the reduction of power losses (Fig. 5-33).

(a)

(b)

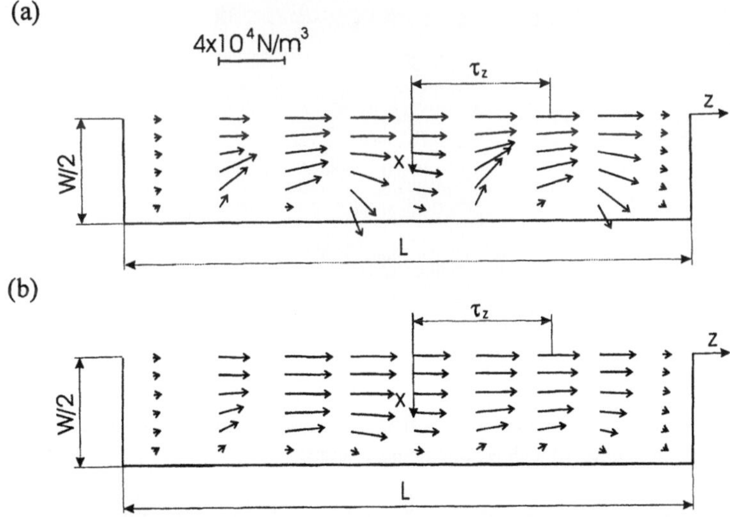

Figure 5-32. Distributions of tangential force component calculated at the channel bottom: (a) without compensating winding, (b) with compensating winding

(a)

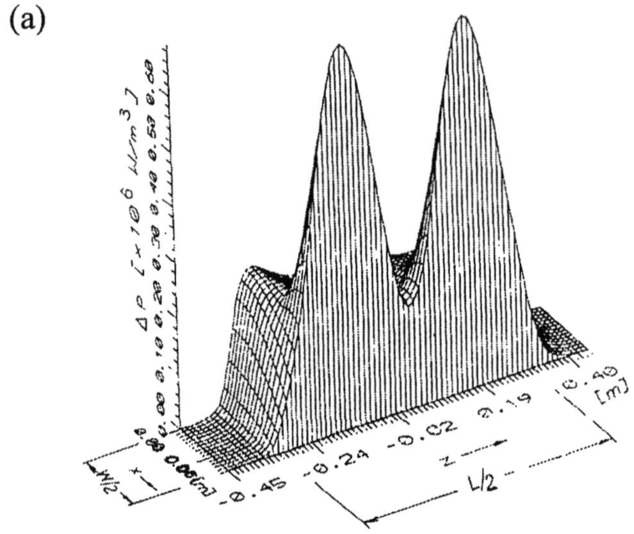

(b)

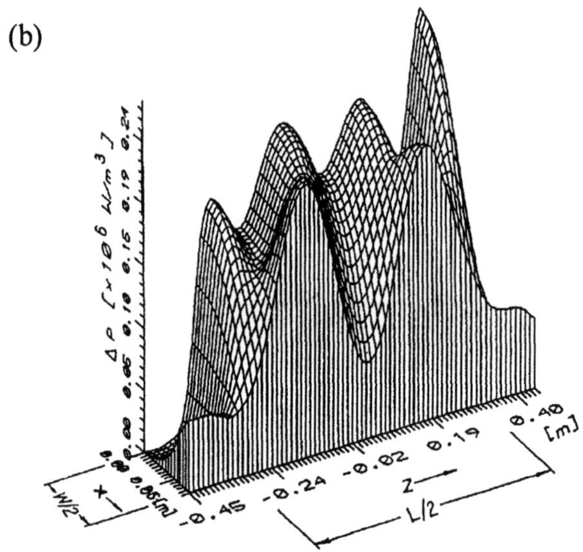

Figure 5-33.   Power loss density distribution at channel bottom: (a) without compensating
winding, (b) with compensating winding

More on applying the Fourier's series technique for the modeling of
liquid metal pumps can be found in [18, 48]. In [19] the calculation are done
when the space and time harmonics of currents are taken into account.

Appendix

I. LINEAR CURRENT DENSITY

In order to include a discrete winding distribution in slots, the current density distribution of pth phase in the current layer of mathematical model presented in Fig. 3-2 is described by the periodic function shown in Fig. I-1.

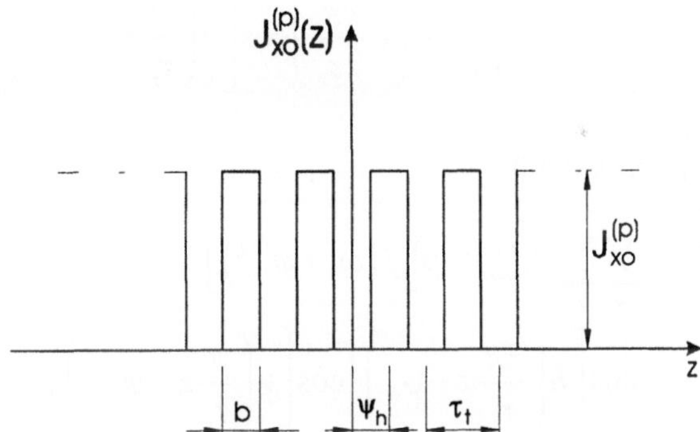

Figure I-1.   Periodic function of the linear current density of pth phase showing discrete winding distribution in slots

It can be represented by the following Fourier's series:

$$J_x^{(p)}(t,z) = J_{xo}^{(p)}(t)\frac{b}{\tau_t} + \sum_h J_{xh}^{(p)}(t)\cos\left[h\left(\frac{2\pi}{\tau_t}z - \psi_h\right)\right] \tag{I.1}$$

where:

$$J_x^{(p)}(t) = 2J_{xo}^{(p)}(t)\frac{1}{h\pi}\sin\left(\frac{b\pi}{\tau_t}\right),\tag{I.2}$$

$$J_{xo}^{(p)}(t) = \frac{I_o^{(p)}w^{(p)}}{b}\cos\left(\omega t + \varphi^{(p)}\right)\quad\text{- linear current density per slot}$$
$$\text{width,}\tag{I.3}$$

w – number of turns per slot,
b – slot opening,
ψ_h – shift of the slot center in relation to co-ordinate system,
$h = 1,2,3,...$
$\varphi^{(p)}$ – phase displacement angle.

Each phase is represented by the current density distributed as shown in Fig. I-2. It can be described in the following way:

$$J_x^{(p)}(t,z) = \sum_v \left\{ J_{xo}^{(p)}(t)\frac{b}{\tau_t} + \sum_h J_{xh}^{(p)}(t)\cos\left[h\left(\frac{2\pi}{\tau_t}z - \psi_h\right)\right]\right\}$$
$$\cdot\frac{4}{v\pi}\sin\left(v\frac{c\pi}{2\tau_z}\right)\cos\left[v\left(\frac{\pi}{\tau_z}z - \psi^{(p)}\right)\right]\tag{I.4}$$

or using complex notation

$$J_x^{(p)}(t,z) = \sum_v \sum_h J_{mxvh}^{(p)} \exp\left[j\left(\omega t + \varphi^{(p)}\right)\right]$$
$$\cdot\cos\left[h\left(\frac{2\pi}{\tau_t}z - \psi_h\right)\right]\cos\left[v\left(\frac{\pi}{\tau_z}z - \psi^{(p)}\right)\right]\tag{I.5}$$

where:

$$J_{mxvh}^{(p)} = \frac{4}{v\pi}\sin\left(v\frac{c\pi}{2\tau_z}\right)J_{mxh}^{(p)},\tag{I.6}$$

$$J_{mxh}^{(p)} = J_{xoh}^{(p)}\frac{b}{\tau_t}\qquad\text{at } h = 0\tag{I.7}$$

$$J_{mxh}^{(p)} = 2J_{xo}^{(p)} \frac{1}{h\pi} \sin\left(h\frac{b\pi}{\tau_t} \right) \qquad \text{at } h = 1,2,3,\ldots \tag{I.8}$$

$v = 1,3,5,\ldots$

c – phase winding group width (see Fig. I-2)

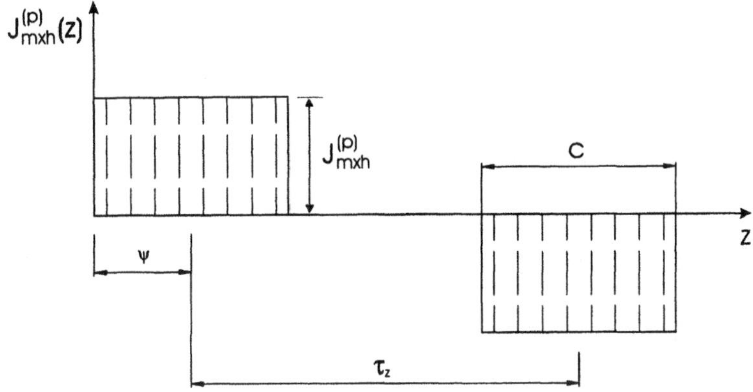

Figure I-2. Periodic function of the linear current density of pth phase showing the phase winding distribution

The linear current density described by Eqn. I.5 can be written as a sum of four traveling waves:

$$J_x^{(p)}(t,z) = \sum_v \sum_h \frac{1}{4} J_{mxvh}^{(p)} \left\{ \begin{array}{l} \exp j\left[\left(\omega t + \varphi^{(p)}\right) - h\left(\frac{2\pi}{\tau_t}z - \psi_h\right) - v\left(\frac{\pi}{\tau_z}z - \psi^{(p)}\right) \right] + \\[2mm] + \exp j\left[\left(\omega t + \varphi^{(p)}\right) + h\left(\frac{2\pi}{\tau_t}z - \psi_h\right) - v\left(\frac{\pi}{\tau_z}z - \psi^{(p)}\right) \right] + \\[2mm] + \exp j\left[\left(\omega t + \varphi^{(p)}\right) - h\left(\frac{2\pi}{\tau_t}z - \psi_h\right) + v\left(\frac{\pi}{\tau_z}z - \psi^{(p)}\right) \right] + \\[2mm] + \exp j\left[\left(\omega t + \varphi^{(p)}\right) + h\left(\frac{2\pi}{\tau_t}z - \psi_h\right) + v\left(\frac{\pi}{\tau_z}z - \psi^{(p)}\right) \right] \end{array} \right\} \tag{I.9}$$

The function I.9 written in a short form is as follows

$$J_x^{(p)}(t,z) = \sum_v \sum_h \sum_q \sum_t J_{mxvhqt}^{(p)} \exp\left[j\left(\omega t - \frac{\pi}{\tau_{zvh}}z \right) \right] \tag{I.10}$$

where

$$J^{(p)}_{mxvhqt} = \frac{1}{4} J^{(p)}_{mxvh} \exp\left[j\left(\varphi^{(p)} + hq\psi_h + vt\psi^{(p)}\right)\right] \tag{I.11}$$

$$\tau_{zvh} = \frac{1}{q\dfrac{2h}{\tau_t} + t\dfrac{v}{\tau_z}}\,, \qquad q,t = 1,-1 \tag{I.12}$$

The total primary current density is a sum of all m-phase current densities

$$J_x(t,z) = \sum_v \sum_h \sum_q \sum_t J_{mxvhqt} \exp\left[j\left(\omega t - \frac{\pi}{\tau_{zvh}} z\right)\right] \tag{I.13}$$

where:

$$J_{mxvhqt} = \sum_{p=1}^{m} J^p_{mxvhqt} \tag{I.14}$$

m – number of primary phases

II. ELECTROMAGNETIC FIELD EQUATIONS

To determine the performance characteristics the most common approach is to calculate them from an equivalent circuit of the motor. The parameters of the circuit need to be determined from the electromagnetic field, which is found from the field model of the motor. As was mentioned in Introduction, to find the field components, the analytical method is used that is based on Fourier's series technique. In this method the motor is represented by the N layer structure shown in Fig. II-1.

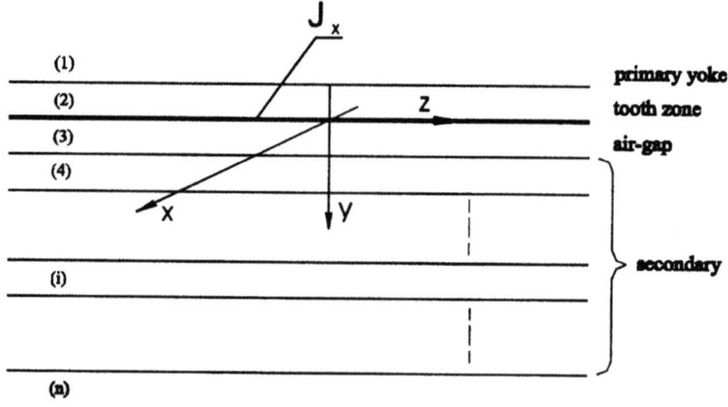

Figure II-1. Multi-layer field model of the IM-2DMF

Each layer is homogeneous, isotropic and linear and defined by magnetic permeability μ_i, conductivity γ_i and velocity v_i. The primary winding is reduced to an infinitely current sheet placed between the second and the third layer. The first layer corresponds to the teeth zone with given saturation, represented by the finite value of permeability μ_2. The current density of the current layer is defined by the function:

$$J = J_m \exp[j(\omega t - \frac{\pi}{\tau_x}x - \frac{\pi}{\tau_z}z)] \tag{II.1}$$

where J_m is the amplitude of linear current density and $\omega = 2\pi f$, is the angular frequency.

In the homogeneous space the electromagnetic field is determined from the Maxwell's equations written in the form:

$$\nabla \times \vec{H} = \vec{J} \tag{II.2}$$

$$\nabla \times \vec{E} = -\frac{\partial \vec{B}}{\partial t} \tag{II.3}$$

$$\nabla \cdot \vec{B} = 0, \qquad \nabla \cdot \vec{D} = 0 \tag{II.4}$$

For the isotropic and the homogeneous environment:

$$\vec{D} = \varepsilon \vec{E}, \qquad \vec{B} = \mu \vec{H} \tag{II.5}$$

Besides, for the moving area

$$\vec{J} = \gamma(\vec{E} + \vec{v} \times \vec{B}) \tag{II.6}$$

When the co-ordinate system is attached to the moving area then frequency is expressed in terms of the slip s_{xz} and the Eqn. II.6 can be reduced to $\vec{J} = \gamma \vec{E}$.

For convenience the vector potential A is used. In the problem discussed here two vector potential components occur A_x and A_z. The transformed Maxwell's equations for the ith layer expressed in the complex number form are as follows:

$$\nabla^2 A_{mxi} = \alpha_i^2 A_{mxi}$$
$$\nabla^2 A_{mzi} = \alpha_i^2 A_{mzi} \tag{II.7}$$

where:

$$\alpha_i^{\,2} = j\omega s_{ixz}\gamma_i\mu_i \qquad\qquad (\text{II.8})$$

The relations between the electric field intensity E and the magnetic flux density B and the vector potential A in the ith layer define the equations:

$$B_{mi} = \Delta \times A_{mi}$$

$$E_{mi} = -j\omega A_{mi} \qquad\qquad (\text{II.9})$$

Since there are no y-directed currents in the computational model, the components E_y and A_y are assumed to be zero. Thus the Eqns. II.9 written for particular components have the form:

$$E_{mxi} = -j\omega A_{mxi}$$

$$E_{mzi} = -j\omega A_{mzi}$$

$$B_{mxi} = \frac{\partial A_{mzi}}{\partial y}$$

$$B_{myi} = \frac{\partial A_{mxi}}{\partial z} - \frac{\partial A_{mzi}}{\partial x} \qquad\qquad (\text{II.10})$$

$$B_{mzi} = \frac{\partial A_{mxi}}{\partial y}$$

The final solutions of the above equations, found using the boundary conditions (H_x, H_z and B_y should be continuous at the boundaries) are as follows:

$$E_{mxi} = K_i'(y)F(x,z)$$

$$E_{mzi} = -\frac{\tau_z}{\tau_x}K_i'(y)F(x,z)$$

$$B_{mxi} = -j\frac{\beta_i}{\omega}\frac{\tau_z}{\tau_x}K_i''(y)F(x,z) \qquad\qquad (\text{II.11})$$

$$B_{myi} = \frac{\alpha_o}{\omega}\frac{\tau_z}{\pi}K_i'(y)F(x,z)$$

$$B_{mzi} = -j\frac{\beta_i}{\omega}K_i''(y)F(x,z)$$

where

$$K_i^{'} = C_i^{'} \exp(\beta_i y_i) + C_i^{''} \exp(-\beta_i y_i)$$

$$K_i^{''} = C_i^{'} \exp(\beta_i y_i) - C_i^{''} \exp(-\beta_i y_i)$$

$$F(x,z) = \exp[-j(\frac{\pi}{\tau_x} x + \frac{\pi}{\tau_z} z)]$$

The constants $C_i^{'}$ and $C_i^{''}$ are expressed in the following way

$$C_i^{'} = \frac{W_i^{'}}{W}, \qquad C_i^{''} = \frac{W_i^{''}}{W}$$

where

$$W_1^{'} = 2 J_x M_2 \cdot M_3^{'}, \qquad W_1^{''} = 0$$

$$W_2^{'} = J_x M_3^{'}(M_2 + 1), \qquad W_2^{''} = J_x M_3^{'}(M_2 - 1),$$

For $i=3,4,...,(n-1)$

$$W_i^{'} = (-2)^{i-2} J_x \exp(-\beta_i g_i)(M_{i+1}^{''} + M_{i+1}^{'})N_2$$

$$W_i^{''} = -(-2)^{i-2} J_x \exp(-\beta_i g_i)(M_{i+1}^{''} - M_{i+1}^{'})N_2$$

and if $i=n$

$$W_n^{'} = 0, \qquad W_n^{''} = (-2)^{n-2} J_x N_2$$

$$W = M_2^{''} - M_2^{'}$$

For $i=3,4,...,(n-1)$

$$M_i^{'} = 2[M_{i+1}^{''} \sinh(\beta_i d_i) - M_{i+1}^{'} \cosh(\beta_i d_i)]$$

$$M_i^{''} = -2[M_{i+1}^{''} \cosh(\beta_i d_i) - M_{i+1}^{'} \sinh(\beta_i d_i)]$$

and for $i=n$

$$M_i^{'} = -\exp(-\beta_n g_{n-1})$$

$$M_i^{''} = M_n \exp(-\beta_n g_{n-1})$$

$$J_x = -j \frac{\omega \mu_2}{4\beta_2} J_{mx}$$

$$N_2 = M_2 \cosh(\beta_2 d_2) + \sinh(\beta_2 d_2)$$

For $i=2,3,...,n$

$$M_i = \frac{\beta_i \mu_{i-1}}{\beta_{i-1} \mu_i}$$

$$g_i = g_{i-1} + d_i \text{ and } g_1 = 0$$

For $i=1,2,...,n$

$$\beta_i^2 = \alpha_o^2 + \alpha_i^2$$

$$\alpha_o^2 = \left(\frac{\pi}{\tau_x}\right)^2 + \left(\frac{\pi}{\tau_z}\right)^2$$

If the current of pth phase is extracted from the linear current density the Eqns. II.11 take the form:

$$E_{mxi}^{(p)} = I^{(p)} K_i^{'(p)}(y) F(x,z)$$

$$E_{mzi}^{(p)} = -\frac{\tau_z}{\tau_x} I^{(p)} K_i^{'(p)}(y) F(x,z)$$

$$B_{mxi}^{(p)} = -j\frac{\beta_i}{\omega}\frac{\tau_z}{\tau_x} I^{(p)} K_i^{''(p)}(y) F(x,z) \qquad\qquad (\text{II.12})$$

$$B_{myi}^{(p)} = \frac{\alpha_o}{\omega}\frac{\tau_z}{\pi} I^{(p)} K_i^{'(p)}(y) F(x,z)$$

$$B_{mzi}^{(p)} = -j\frac{\beta_i}{\omega} I^{(p)} K_i^{''(p)}(y) F(x,z)$$

Constants $K_i^{'}$ and K_i^{*} do not differ from the ones derived above, except the linear current density J_x, which does not contain current $I^{(p)}$.

References

1. Bevan R.J., Kalman G.: Non-Uniform Power Distribution in Linear Induction Motors Due to End Effects, IEEE Trans. on PAS., Vol. PAS-98, No. 5, 1979, pp. 1516-1521.

2. Cathey J.J.: Helical Motion Induction Motor, Proc. IEE, vol. 132, pt. B, No. 2, 1985, pp. 112-114.

3. Cathey J.J. and Rabiee M.: Verification of an Equivalent Circuit Model for a Helical Motion Induction Motor, IEEE Trans. On Energy Conversion, vol. EC, 1988 pp. 660–666

4. Chalmers B.J.: The Torus Generator, Manchester Machines Seminars 1993, pp.98-106.

5. Dabrowski M., Gieras J.: Induction Machines with Solid Rotor, PWN Warszawa-Poznan, 1977 (in Polish).

6. Dukowicz J.: Analysis of Linear Induction Machines with Discrete Windings and Finite Iron Length, IEEE Trans. on PAS, Vol. PAS-96, No. 1, 1977, pp. 66-73.

7. Easham J.F., Alwash J.H.: Transverse-Flux Tubular Motors, Proc. IEE, Vol. 119, No. 12, 1972, pp. 1709-1718.

8. Eliot D.G.: Matrix Analysis of Linear Induction Machines, Report for U.S. Department of Transportation, No. FRA-ORD-75-77, Sept. 1975.

9. Fikus F. and Wieczorek T.: *Magnetohydrodynamic devices in foundries and steel mills*, Silesia, Katowice, 1979 (in Polish)

10. Fleszar J.: Analysis of Operation of Rotary-Linear Induction Motor, Ph.D. Dissertation, Lodz University of Technology 1981.

11. Fleszar J., Mendrela E.: Twin-Armature Rotary-Linear Induction Motor, IEE Proc. Part B, Vol. 130, No. 3, 1983, pp. 186-192.

12. Fleszar J., Gierczak E., Mendrela E.: Calculation of electromagnetic field and electromechanical parameters of disc-type induction motor with double sided stator and twin solid rotors, Archives of Electrical Engineering 2000, No 2, pp. 159-170

13. Fleszar J., Gierczak E., Mendrela E., Evaluation of equivalent circuit parameters and electromechanical characteristics of unsymmetrical disc-type induction motor with double sided stator and twin solid rotors, Archives of Electrical Engineering 2001, No 1, pp. 31-42

14. Foggia A., Sabonnadiere J.C., Silvester P.: Finite Element Solution of Saturated Travelling Magnetic Field Problems, IEEE Trans. on PAS, Vol. PAS-94, No. 3, 1975, pp. 866-781.

15. Gieras J.F.: *Linear Induction Drives*, Oxford University Press, 1998
16. Gierczak E.: Analysis of Electromagnetic Field and Forces in Rotary-Linear Induction Motors, Ph.D. Dissertation, Lodz University of Technology 1980.
17. Gierczak E., Mendrela E.A.: Influence of design parameters on LIM performance, Archiwum Elektrotechniki, No. 2, 1994, pp. 393-402
18. Gierczak E., Mendrela E.A., Mendrela E.M.: Calculation of an electromagnetic field in a three-phase induction pump with a compensating winding, IEEE Trans. on Magnetics, Vol.29, No.5, 1993, pp. 2289-2294.
19. Gierczak E., Mendrela E.A.: Three Dimensional-Space Analysis of the Electromagnetic Field in Induction Pump with Nonuniform Distribution of Molten Metal Velocity and Nonsinusoidal Current, Archiv fur Elektrotechnik 73, 1990, pp.245- 252
20. Gierczak E. and Mendrela E.A.: Magnetic flux, current, force and power loss distribution in twin-armature rotary-linear induction motor, Scientia Electrica, Vo. 31, pp. 65-74, 1985
21. Gerrard J., Paul. R.A.: Rectilinear Screw-Thread Reluctance Motor, Proc. IEE, No. 11, 1971, pp. 1575-1584.
22. Jeon W.J., Tanabiki M., Onuki T. Yoo, J.Y.: Rotary-Linear Induction Motor Composed of Four Primaries with Independently Energized Ring-Windings, Industry Applications Conference, IAS'97, Conference Record of the 1997 IEEE, Vol. 1
23. Kaminski G.: Magnetic Field and Performance Asynchronous Motor with Linear-rotary Motion, Rozprawy Elektrotechniczne, Z. 4, 1980, pp. 963-970. (in Polish)
24. Kaminski G., Kant M.: Electric Motor with Complex Motion of Spherical Rotor: Przeglad Elektrotechniczny, Z. 8-7, 1981, pp. 278-280. (in Polish)
25. Kaminski G.: *Electrical Complex Motion Motors*, Oficyna Wydawnicza Politechniki Warszawskiej, Warszawa 1994, (in Polish)
26. Kaminski G., Smak A.: Modern Constructions of Motors with Spherical Rotors, Archives of Electrical Engineering,Vol. L, No 197-3/2001, pp 215-233
27. Kant M.: Hybrid Linear-Rotating Motor, Proc. IEE, Vol. 124, No. 5, 1977.
28. Laitwaite E.R.: *Propulsion without Wheels*, Eng. Univ. Press, London 1966.
29. Laitwaite E.R.: *Transport without Wheels*, Elec. Science, London 1979.
30. Laitwaite E.R.: *Induction Machines for Special Purposes*, George Newnes, London 1966.
31. Lee C.H., Chin C.Y.: A Theoretical Analysis of Linear Induction Motors, IEEE Trans. On PAS, Vol. PAS-98, No. 2, 1979, pp. 679-688.
32. Mendrela E.: Rotary-Linear Induction Motor, Ph.D. Dissertation, Wroclaw University of Technology 1975. (in Polish)
33. Mendrela E.: Rotary-Linear Induction Motor, Rozprawy Elektrotechniczne, t. 22, z. 2, 1976, pp. 383-406. (in Polish)
34. Mendrela E.: Analitycal Method of determination of electromagnetic field and forces in Twin-Armature Rotary-Linear Induction Motor with Shielded Rotor, Rozprawy Elektrotechniczne, t. 25, z. 2, 1979, pp. 385-400. (in Polish)
35. Mendrela E., Kaplon A.: Rotary-Linear Induction Motor with Rotating-Traveling Field, Proc. of ICEM, part. III, Budapest 1982.
36. Mendrela E., Kaplon A., Muszczak W.: Test Stand for Rotary-Linear Motors, Zesz. Nauk. Polit. Swietokrzyskiej, Elektryka 7, 1980, pp. 121-129. (in Polish)
37. Mendrela E., Fleszar J., Gierczak E.: Method of Determination of the Distance Between Fictitious Primaries in Computational Model of Linear Induction Motor Used in Fourier Series Technique, Archive fur Elektrotechnik No. 66, 1983, pp. 151-156.
38. Mendrela E., Gierczak E.: Two-Dimensional Analysis of Linear Induction Motor Using Fourier's Series Method, Archiv fur Elektrotechnik No. 65, 1982, pp. 97-106.

39. Mendrela E., Gierczak E.: Calculation of Transverse Edge Effects of Linear Induction Motor Using Fourier's Series Method, Archiv fur Elektrotechnik No. 65, 1982, pp. 161-165.

40. Mendrela E.A.: *Induction Motor with Two Degrees of Mechanical Freedom (An Analysis of Electromagnetic Phenomena and Electromechanical Properties at Steady State Conditions)*, Elektryka 14, Technical University of Kielce, 1984 (in Polish)

41. Mendrela E.A., Gierczak E.: Double-Winding Rotary-Linear Induction Motor, IEEE Trans. on Energy Conversion, EC-2, March 1987, pp.47-54

42. Mendrela E.A., Gierczak E.: Influence of Primary Twist on Linear Induction Motor Performance, IEEE Trans. on Magnetics, MAG-21, Nov. 1985, pp.2664-2671

43. Mendrela E.A.: Steady-State Stability Conditions of Rotary-Linear Electric Motors, Scientia Electrica 31, 1985, pp.125-139

44. Mendrela E.A., Gierczak E.: Performance of Rotary-Linear Induction Motor with Rotating-Traveling Field, Electric Machines and Power System, No.9, 1984, pp.171-178

45. Mendrela E.A., Fleszar J.: Equivalent Circuit of Induction Motors with Two Degrees of Mechanical Freedom, Acta Technica CSAV, No.6, 1983, pp.697-715

46. Mendrela E.A., Fleszar J.: Performance characteristics of twin-armature rotary-linear induction motor, Acta Technica CSAV, No.5, 1983, pp.595-603

47. Mendrela E.A., Mendrela E.M., Brol L.: Evaluation of magnetic field and forces in single-phase induction disc motor with multi-layer rotor, Archives of Electrical Engineering, No.2,1995, pp. 271-281.

48. Mendrela E.A., Mendrela E.M.: Three dimensional space analysis of the electromagnetic field in induction pump with nonuniform distribution of molten metal velocity, Archiwum Elektrotechniki, Vol.3, 1994, pp.445-465

49. Mendrela E.A., Macek-Kamińska K.: Disc induction motor with double-sided stator and twin rotors, Archives of Electrical Engineering, 1999, No 1-2,pp. 51-62

50. May H., Mosebach H., Weh H.: Numerical Treatment of Transverse Edge Effects in Linear Induction Motors, Electric Machines and Electromechanics 4, 1979, pp. 321-330.

51. Mosebach H., Huhns T., Pierson E.S., Herman D.: Finite Length Effects in Linear Induction Machines with Different Iron Contours, IEEE Trans. On PAS-96, 1977, pp. 1087-1093.

52. Nasar S.A., Del Cid L.: *Linear-Motion Electric Machines*, New York, J. Wiley and Sons, 1976.

53. North G.G.: Harmonic Analysis of a Short Stator Linear Induction Motor Using a Transformation Technique, IEEE Trans. on PAS, Vol. PAS-92, No. 5, 1973, pp. 1733-1743.

54. Onuki T., Jeon W.J., Tanabiki, M.: Induction Motor with Helical Motion by Phase Control, IEEE Trans. on Magnetics, Vol. 33, Issue: 5, Part: 2, 1997 pp. 4218 –4220

55. Oset A.: Electromagnetic Field Analysis in Rotary-Linear Induction Motor Taking into Account End Effects, Ph.D. Dissertation, Warsaw University of Technology, 1982. (in Polish)

56. Pierson E.S., Hanitsch R., Huhns T., Mosebach H.: Predicted and Measured Finite-Width-Effects in Linear Induction Machines, IEEE Trans. on PAS, Vol. PAS-96, 1977, pp. 1081-1086.

57. Poloujadoff M., Morel B., Bolopion A.: Simultaneous Consideration of Finite Length and Finite Width of Linear Induction Motors, IEEE Trans. on PAS, Vol. PAS-99, No. 3, 1980, pp. 1172-1179.

58. Poloujadoff M.: *The Theory of Linear Induction Machinery*, Oxford University Press, 1980.

59. Rabiee M., Cathey J.J.: Verification of a Field Theory Analysis Applied to a Helical Motion Induction Motor, IEEE Trans. on Magnetics, vol. MAG, 1988, pp. 2125–2132

60. Rabiee M.: Analysis and Control of Helical Motion Induction Motor, Ph.D. Dissertation, University of Kentucky, Lexington, 1987.
61. Turowski J.: *Technical Electrodynamics*, WNT Warsaw, 1968. (in Polish)
62. Turowski J.: *Electromagnetic Calculations of Machine Elements and Electromechanics*. WNT Warsaw, 1982. (in Polish)
63. Yamamura S.: *Theory of Induction Motors*, New York, John Wiley and Sons, 1972.
64. Zawadzki A., Gierczak E., Mendrela E.A.: Calculation Method of Linear Induction Motor Supplied from Symmetrical Voltage Source, Archiv fur Elektrotechnik 71, 1988, pp. 221-227.

Index

MIX
Papier aus verantwortungsvollen Quellen
Paper from responsible sources
FSC® C105338

If you have any concerns about our products,
you can contact us on
ProductSafety@springernature.com

In case Publisher is established outside the EU,
the EU authorized representative is:
**Springer Nature Customer Service Center GmbH
Europaplatz 3, 69115 Heidelberg, Germany**

Printed by Libri Plureos GmbH
in Hamburg, Germany